Civilians into soldiers

Manchester University Press

Cultural History of Modern War

Series editors
Ana Carden-Coyne, Peter Gatrell, Max Jones, Penny Summerfield and
Bertrand Taithe

Already published

Julie Anderson   *War, disability and rehabilitation in Britain: soul of a nation*
Rachel Duffett   *The stomach for fighting: food and the soldiers of the First World War*
Christine E. Hallett   *Containing trauma: nursing work in the First World War*
Jo Laycock   *Imagining Armenia: Orientalism, ambiguity and intervention*
Chris Millington   *From victory to Vichy: veterans in inter-war France*
Juliette Pattinson   *Behind enemy lines: gender, passing and the Special Operations Executive in the Second World War*
Chris Pearson   *Mobilizing nature: The environmental history of war and militarization in Modern France*
Jeffrey S. Reznick   *Healing the nation: soldiers and the culture of caregiving in Britain during the Great War*
Jeffrey S. Reznick   *John Galsworthy and disabled soldiers of the Great War: with an illustrated selection of his writings*
Michael Roper   *The secret battle: emotional survival in the Great War*
Penny Summerfield and Corinna Peniston-Bird   *Contesting home defence: men, women and the Home Guard in the Second World War*
Wendy Ugolini   *Experiencing war as the 'enemy other': Italian Scottish experience in World War II*
Colette Wilson   *Paris and the Commune, 1871–78: the politics of forgetting*
Laura Ugolini   *Civvies: middle-class men on the English Home Front, 1914–18*
http://www.arts.manchester.ac.uk/subjectareas/history/research/cchw/

Civilians into soldiers

War, the body and British Army recruits, 1939–45

~

EMMA NEWLANDS

Manchester University Press

Copyright © Emma Newlands 2014

The right of Emma Newlands to be identified as the author of this work has been asserted by her in accordance with the Copyright, Designs and Patents Act 1988.

Published by Manchester University Press
Altrincham Street, Manchester M1 7JA, UK
www.manchesteruniversitypress.co.uk

British Library Cataloguing-in-Publication Data is available

ISBN 978 0 7190 8804 9 hardback

First published 2014

The publisher has no responsibility for the persistence or accuracy of URLs for any external or third-party internet websites referred to in this book, and does not guarantee that any content on such websites is, or will remain, accurate or appropriate.

Typeset by JCS Publishing Services Ltd, www.jcs-publishing.co.uk

For Grace

Contents

List of illustrations	*page* viii
List of tables	ix
Acknowledgements	xi
Abbreviations	xiii
Introduction	1
1 Examination	26
2 Training	53
3 Experimentation	90
4 Active service	116
5 Fear, wounding and death	154
Conclusion	184
Bibliography	191
Index	211

Illustrations

1. A prospective recruit undergoes a medical examination at the drill hall in the Old Dukes Road, Euston (Imperial War Museums, London H 38584)   *page* 29
2. Infantry recruits undergoing rifle drill at Chichester Barracks, 1939 (Imperial War Museums, London H 521)   67
3. 'Just another "Claptrap". You can ruin your future with – V.D!' Poster designed by Stacey Hopper to warn Allied troops in Italy about the dangers of venereal disease, 1943–44 (Wellcome Library, London L0023434)   134
4. Men of the Norfolk Regiment receive their rum ration before going out on patrol in France in January 1940 (Imperial War Museums, London F 2264)   157

Tables

1. Medical classifications of soldiers by categories, February 1940   *page* 30
2. Court martial convictions, British other ranks overseas, 1 September 1939–31 August 1945   138
3. Weekly rates of disablement pension (s.d.) for non-regular soldiers, by rank and degree of disablement, September 1939   171
4. Yearly rates of disablement pension (£) for non-regular officers, by rank and degree of disablement, September 1939   171

Acknowledgements

This book is about the place of the body in British military life and culture during the Second World War. My interest in this was fostered during my postgraduate studies at the University of Strathclyde. I am deeply indebted to my supervisors Professor James Mills and Professor Arthur McIvor for their constant support and encouragement. I would also like to thank Dr Juliette Pattinson and Professor Mark Harrison for the feedback and advice they have given me on this work.

I am grateful to the staff at The National Archives in Kew and at the Wellcome Library in London, particularly Ross MacFarlane, for helping me to locate military medical records. Staff at the Imperial War Museums Sound Archive have also been unfailingly helpful. Without them I would not have been able to access so many of the personal testimonies of Second World War veterans that have been crucial to this study. Particular thanks go to Richard Hughes, who contacted the families of many of the men whose stories are included in this work. To my editor at Manchester University Press, Emma Brennan, thank you for your patience and guidance. Thanks also go to my good friend Guy Taylor for his help and advice.

The research on which this book is based would also have been so much harder without the financial assistance I received from the Centre for the Social History of Health and Healthcare at Strathclyde and Glasgow Caledonian Universities. Without this I would not have been able to make many of the research trips that were so important to this work.

On a more personal note, I would like to thank all of my colleagues at the University of Strathclyde and the Centre for the Social History of Health and Healthcare. Particular thanks go to Matt Smith, who was never too busy to read my work and provide insightful comments. Special

Acknowledgements

thanks also go to Linsey Robb, who has been a kind and generous friend and has shared in my highs and lows throughout the writing process. My heartfelt thanks go to my parents, for all the ways that they have helped me over the years. They have been a constant source of encouragement and support. I thank my husband Stuart for his unfaltering belief in me. Finally, I dedicate this book to my daughter Grace, who was born during the months that it was written. Without her loving attention it would have been finished in half the time.

Abbreviations

ACI	Army Council Instruction
AFHQ	Army Forces Headquarters
AMGOT	Allied Military Government of Occupied Territories
ATS	Auxiliary Territorial Service
AWOL	absent without leave
CAB	Cabinet Office
CSM	Command Sergeant Major
FD	Records of the Medical Research Council
IWM	Imperial War Museums
L of C	Lines of Communication
MO	Medical Officer
MOA	Mass Observation Archive
MOD	Ministry of Defence
MRC	Medical Research Council
NAAFI	Navy, Army and Air Force Institute
NCO	Non-Commissioned Officer
POW	prisoner of war
PT	physical training
RAMC	Royal Army Medical Corps
RAP	regimental aid post
RSM	regimental sergeant major
TC	Topic Collection (in Mass Observation Archive)
TNA	The National Archives
TS	Treasury Solicitor
WAAF	Women's Auxiliary Air Force
YMCA	Young Men's Christian Association

Introduction

In 1942 nineteen-year-old Roy Bolton arrived for basic training at Richmond Barracks in London. Here he was issued with a uniform, given a haircut and subjected to a regime consisting of weapons instruction, physical training and drill. He later recalled in an interview:

> I didn't think it was very nice at all. It was difficult. I didn't take to it at all well because in those days anyway I was somewhat clumsy I think, in a sort of bodily way. I found the marching and even keeping step, not too difficult keeping step, but not entirely easy, and then the sudden changes in direction, the right turns, the left turns, the about turns, these I did find tricky. Occasionally I distinguished myself by marching off in the wrong direction.[1]

Roy was one of almost three million men recruited into the ranks of the British Army during the Second World War, making it by far the largest of the three armed services. The majority of these men, like Roy, were conscripts, recruited under the government's National Service Acts.[2] Roy's story illustrates how the body can be used as a lens through which to understand the history of that conscript army. The most obvious point seems to be that the body was at the heart of the experience of military service. Long before he was armed and sent into conflict, Roy was subjected to a regime of physical interventions by the military authorities. His head was shaved, he was issued with new clothes and he was forced to exercise in time with other men in a dedicated space, the barrack square, selected for the purpose. He was no longer able to wear his hair as he chose, to choose clothes that he preferred, or to employ or to rest his body as he saw fit. Clearly, his body was a key concern for the military authorities in the Second World War. Their ambitions for it and the techniques that they employed in order to achieve these will be examined in this book.

It is clear from Roy's account, however, that whatever the designs of the military authorities, these were not always successfully realised. This resistance to the army's orders was expressed through the body; Roy would miss turns and head off in the wrong direction and was literally out of step with his fellow recruits. Moreover, this resistance to military designs was blamed on his body. Roy claimed that he was 'somewhat clumsy' in his own recollection of the period, suggesting that although he wished to comply with the army's orders his physiological make-up prevented him from doing so. Questions about how far Roy was resisting the military authorities when it was his body rather than his will that seemed to be responsible for his failure to comply, and how far the body could act as a tool of resistance, seem to promise much for historians seeking to explore the place of the human body in history.

Yet, while the brief extract above can be read as a story of the failure of the army and Roy to order his body in line with the demands of the military, it also implies that this was unusual. In other words, the very fact that Roy recalled his failure – and the fact that his failure seemed to make him stand out from the others – suggests that the army was successful in imposing its control. The reasons for this success are as important as the causes of the occasional failures, as they also promise to engage with debates about the place of the body in history. The extent to which compliance with regimes of corporeal transformation imposed by modern institutions was unthinking, and how far it was an act of human agency, would seem to be important for understanding modernity itself.

Finally, Roy's story above is provided from the testimony of the soldier himself. Oral evidence is central to this work as it seeks to engage with the lived experiences of wartime recruits. Military records, medical documents, government records and other sources are important for exploring the designs of the military authorities and the techniques that they used to achieve them. However, it is the voices of the men themselves that promise to reveal how far bodies were transformed by the demands of the army, and the reasons for this.

The body

The theoretical orientation of this study is grounded in the social and cultural understandings of the body that have emerged in recent decades. Much of this has been influenced by the work of Michel Foucault, who suggested that the body was both the 'object' and 'target' of power within institutions of the modern state – the school, factory, prison

Introduction

and military barracks – which rendered bodies 'docile'.³ Describing the eighteenth-century French Army, Foucault wrote that 'The soldier has become something that can be made; out of a formless clay, an inapt body, the machine required can be constructed; posture is gradually corrected; a calculated constraint runs slowly through each part of the body, mastering it, making it pliable, ready at all times, turning silently into the automatism of habit; in short, one has "got rid of the peasant" and given him "the air of the soldier".'⁴ Foucault claimed that it is not necessarily through oppressive action that this control over individuals is achieved. Rather, the modern subject is produced from two techniques or 'technologies': the 'technologies of power' and the 'technologies of the self'.⁵ Mediating between these two is the 'art of government' or 'governmentality', which allows individuals to become regulated from the 'inside'.⁶

Foucault's analyses have, however, been criticised for failing to recognise the materiality of the body; more recent studies have emerged which examine how the body is constructed within modern society, taking more seriously its existence as a corporeal phenomenon. Chris Shilling suggests that within Foucault's work the body is something of an 'absent presence' that 'vanishes' as a biological entity. Rather than viewing the body as simply the product of power or knowledge, he conceptualises the body as an unfinished biological and social phenomenon that can be transformed as a result of entry into society.⁷ R.W. Connell likewise states that 'bodies, in their own right as bodies, do matter. They age, get sick, enjoy, engender, give birth. There is an irreducible bodily dimension in experience and practice; the sweat cannot be excluded.'⁸ Connell's groundbreaking work on masculinities highlights the importance of this sense of physicality to the cultural interpretation of gender: 'to be an adult male is distinctly to occupy space, to have a physical presence in the world.' While social relations are thus embedded in certain performances, these activities are themselves 'bodily'.⁹ In his book, *Regulating Bodies*, Bryan Turner also conceives of the body as a 'potentiality' which is elaborated by culture and developed by society.¹⁰ Focusing specifically on the role of medicine, which he argues has taken over the moral regulatory functions once occupied by religion, Turner suggests that in the twentieth century bodies have become increasingly rationalised through regimens such as diet and training. One of the key features of this rationalisation of the modern body has been 'medicalisation': the application of scientific knowledge and practice to the production of 'healthy, reliable, effective and efficient bodies'.¹¹

Civilians into Soldiers engages with these ideas about the body to analyse how the British prepared men for military service during the Second World War. It examines the designs of the state, military and medical authorities for the male body in wartime and the techniques that they employed to mobilise, control, treat and transform the bodies that they were presented with. It explores the qualities that were considered necessary, the skills that were taught and the strategies that were used in order to induce bodies to discipline themselves. In doing so, this work uncovers the aims and operations of the wartime state as it sought to organise its citizens for warfare and considers the extent to which the army relied on compliance in order to achieve its objectives.

This book is not, however, only concerned with the ways in which military values were inculcated among wartime recruits. A key objective of this work is to understand the subjective or 'embodied' experiences of soldiers themselves. It considers the feelings that men had towards their own bodies, the ways in which they perceived and used their bodies, and how they responded to the army's efforts to shape and control them. In doing so, it examines the relationship between the body and the self, as men faced new and often challenging demands through their experiences of military life. Numerous case studies of race, age, gender and national identity have highlighted the importance of the body to the construction of the 'self'.[12] Notably, Pierre Bourdieu suggested that class identities are literally and metaphorically embodied. Using the concept of the 'habitus', he argued that individuals acquire certain ways of looking at things according to their position in social space. These dispositions are expressed in the most natural features of the body, such as height, weight and volume. The ways that people treat and use their bodies are thus directly related to their social class.[13] Erving Goffman also suggested that the control of one's own body is fundamental to the maintenance of self-identity. He recognised bodies as material resources, owned by individuals who control and monitor them in ways that allow them to participate in everyday life. Goffman argued that, although bodies themselves are not produced by social forces, the meanings attached to them are determined by 'shared vocabularies of body idiom' – conventionalised forms of non-verbal communication, such as dress, bearing, movements and position, which are outside the immediate control of individuals. Intervening successfully in social life therefore requires a high degree of competency in observing these rules. When individuals do not conform they are assigned a marginalised position within society, leading to a 'spoiled self-identity'.[14]

Introduction

While taking seriously the body as a ground of human experience, both Goffman and Bourdieu nevertheless continued to locate its significance within a classificatory system that exists somehow independently from it. Although individuals actively use and monitor their bodies in order to achieve a particular sense of identity, their behaviours are still governed by a set of dispositions that are imposed from 'outside'. Other scholars have proposed that that embodied behaviour actually shapes society and culture. Drew Leder refers to the human body as an 'intending entity' that is 'not just one thing in the world but a way in which the world comes to be'. He suggests that it is through our bodies that we respond and give meaning to the world around us.[15] Maurice Merleau-Ponty developed the term 'body-subject' to describe the body's actively engaged role within the society.[16] He also recognised human behaviour as having a social and historical base that was drawn from social stock or 'habitus'. However, Merleau-Ponty argued that embodied action is intelligent and purposeful, taking up these skills and techniques and deploying them as and when is appropriate. Through our bodily behaviours we thus assume a position in, and sustain, the world around us.[17] From this perspective, actions which may appear as evidence of compliance or coercion can be read as signs of agency, as individuals manage their bodies in ways that conform to dominant codes and norms in order to achieve productive ends of their own.[18] These theories of embodied agency are central to this book. *Civilians into Soldiers* examines how men used their bodies to participate in military life and how they worked on and invested in their own bodies to fulfil their own goals.

The concept of resistance is also one that will recur throughout this book. The body's capacity to disrupt established codes and norms has been the focus of much scholarly work. Connell, for instance, uses the term 'body-reflexive practices' to describe how bodies, by entering into society, have the ability to change the very relations in which they are engulfed.[19] Indeed, many studies of contemporary military culture highlight the resistance expressed by service personnel through bodily channels. In his work on modern-day infantry recruits John Hockey argues that during basic training, where military discipline is at its most extreme, men respond with various 'corporeal tactics' that allow them to counter the regulation of their bodies in both real and symbolic terms.[20] These include the deliberate misinterpretation of commands when being drilled by a corporal: a very public form of deviance that is open to the gaze of superior ranks.[21] Paul Higate's study of the Royal Air Force also demonstrates the importance of the body in forms of resistance among personnel administrators or 'clerks'. Higate argues that these roles provide a limited outlet for the traditional 'man-of-action' ideal when compared with the combat fighter. As men who occupy the lower reaches of the trade hierarchy, clerks show resistance to their position of subjection through 'embodied coping strategies'.[22] These include participation in 'rumbustious' risk-taking behaviours within the office, such as play fighting, and a commitment to working-out in the gym outside office hours. According to Higate, these activities allow clerks to assert their masculinity and reaffirm the man-of-action image.[23]

While uncovering such open or obvious forms of opposition, *Civilians into Soldiers* also seeks to explore in more detail the complex nature of resistance among army personnel in the Second World War. Sociological studies show that the concept of resistance is inherently problematic. This is because of two key issues: recognition and intent. Jocelyn Hollander and Rachel Einwohner ask, 'must oppositional action be readily apparent to others?' and 'must the actor be aware that she or he is resisting some exercise of power – and intending to do so – for an action to qualify as resistance?'[24] Certain behaviours may, for instance, be interpreted differently in different cultures. What one person recognises as resistance may not be thought of as such by another. Uncovering the motivations behind human behaviour can also be difficult, particularly as individuals can conceal their intention to resist. In this respect, James C. Scott developed the concept of the 'hidden transcript' to describe subtle forms of everyday non-cooperation that can only be safely performed 'offstage'. Scott argues that it is a survival skill of the subordinated to fulfil the expectations of those in charge, producing a more or less credible performance, out of fear or a desire to curry favour. Beyond these public performances, however, individuals behave in ways that contradict established rules and norms but, unlike outright acts of defiance, do not have ominous consequences.[25] Scott suggests, moreover, that these hidden transgressions do not just involve the subordinated. Rather, 'the powerful have their own compelling reasons for adopting a mask.' Resistance in this respect becomes legitimised by the elite.[26] Other writers suggest that resistance does not even need to be intentional. Natalie Armstrong and Elizabeth Murphy emphasise that 'it is important to distinguish between resistance at the behavioural level (e.g. refusal to accept a particular recommended procedure) and resistance at a conceptual level (e.g. rejection of the discourse within which a particular procedure is embedded.'[27] People who wear unconventional styles of dress, for example, do not necessarily intend to assert their deviance.[28]

Introduction

These ideas about resistance are even more important when we are looking at the body because humans are often constrained by their physical limitations. Nick Crossely suggests that physical training presupposes certain abilities of the body. Corporeal routines and skills can, therefore, be immensely transformed only if there is some effective basis to work with.[29] This means that even if we want to conform, we may be prevented from doing so, because our body can betray us and become something 'other' to the self.[30] *Civilians into Soldiers* applies all of these ideas about resistance to the British Army of the Second World War. It examines a wide range of subversive bodily cultures, including those that were individual, collective and institutional. I argue that while some men did come to internalise the control and regulation of their bodies, others simply played at or performed their new military roles, finding alternative safe spaces in which they could pursue their own agendas. This study also draws attention to moments of unintentional resistance, where men were simply unable to cope with the physical demands of military life. As such, this work demonstrates that although the army may have tried to transform civilians into soldiers by using bodily techniques, it was not always able to achieve its objectives. It determines what led to successful outcomes and shows what can be drawn from these instances by way of larger observations about human agency in general.

The British Army, the body and society, 1939–45

There is a wealth of publications on the British Army of the Second World War, including studies of grand campaigns such as Burma, D-Day and the African Desert, histories of individual battalions and recollections and biographies of individual soldiers and senior commanders.[31] The majority of these are, however, popular rather than academic histories. While they do at times convey the feelings, expectations and motivations of men who served, these appear intermittently in what are really chronological narratives. They do not attempt to relate these military experiences to wider historiographical debates about the nature of British society during the war. The scholarly studies of the wartime army that do exist are predominantly works of military history that focus on battle tactics, doctrine, leadership and internal organisation.[32] These are concerned only with top-down institutional arrangements. Their focus has remained on one key issue: military efficiency.

In recent years a minority of historians have turned their attention away from this traditional military approach to consider the social content and character of the British Army during between 1939 and 1945. Notably, Jeremy Crang has explored the nature of the army within Britain as a social institution. Through a series of case studies of officer and other rank selection, officer–man relations, welfare and education, he shows how the military authorities dealt with the problems posed by their new civilian intake, and the efforts that were made to integrate these citizen soldiers into the ranks.[33] In 1942, for example, the General Service Selection Scheme introduced standardised basic recruit training and scientific testing in order to reduce the occurrence of 'square pegs being placed in round holes'.[34] Officer training also came to place more emphasis on 'man management' as a means of improving morale among conscript troops, many of whom looked with scepticism upon the privilege of rank.[35] Ultimately, however, Crang argues that these wartime reforms had limited impact upon the army as a social institution. While there were measureable improvements in efficiency, significant resistance from middle-ranking officers, many of whom were part of the pre-war regular army, meant there was little profound social change. Crang concludes that 'the people's war might have brought the army and nation closer together…but in many ways the army remained a nation apart.'[36]

Focusing more on the combat performance of troops in the field, David French has also examined the military implications of the social composition of the army during these years. Countering the idea that British success stemmed from brute force, he argues that there were three elements that sustained fighting ability: 'the conceptual, the material and the moral'.[37] French suggests, for instance, that British successes after 1942 were not due to a superiority of material resources, since individual weapons remained qualitatively inferior to those of the Germans. Rather, the British developed a growing appreciation of how to use the weapons available to their maximum advantage. In reality, British commanders also failed to take risks and were reluctant to take unnecessary casualties, which they could ill afford. French asserts that 'the British never believed that they could win their battles by pitting man against man, indeed they never believed that they should even try to do so.' In the end, 'it was better to be soldiers than warriors.'[38]

Civilians into Soldiers builds on this scholarship by using the military body as a case study. By placing at the centre of the analysis examinations of physical appearance, aesthetics and the adornment and manipulation of the soldier's body, it uncovers the aims and operations of the army and the wartime state as they sought to prepare male civilians for warfare. Just as importantly, it places these ideas and practices within the wider social and cultural context of mid-twentieth-century Britain. Moreover, this work draws upon the experiences of recruits themselves to understand how modern warfare was seen and represented by ordinary servicemen. Through an exploration of their personal narratives, it examines the investments that men made in their own bodies in order to construct themselves both as citizens and as soldiers. It considers how they too mediated the specialised army environment and wider culture in order to make sense of their new military lives. As such, this book brings together and extends several diverse historiographies, including military history, medical history and the social and cultural history of Britain in the Second World War and the early twentieth century as a whole.

Firstly, *Civilians into Soldiers* adds to existing understandings of the relationship between war and medicine. In his detailed study of the Second World War, Mark Harrison argues that superior medical arrangements in the field gave the British a crucial edge. Better sanitary and hygiene facilities improved the rate of return of casualties to active service and reduced wastage from disease. This, Harrison suggests, stemmed from a unique 'medical consciousness' among officers, who readily embraced modern ideas of social hygiene. Influenced by wider processes of rationalisation within industrial society, they came to see that military success depended on keeping their soldiers fit for service.[39] Julie Anderson has considered what happened to injured and disabled bodies once they left the battlefield and the hospital, suggesting that they can tell us more about the nature of British medical practice during the twentieth century. Anderson argues that the experience of the Second World War was key to the development of the rehabilitation movement. The government's emphasis on ensuring that disabled men re-entered the workplace quickly meant that a 'modern, organised system of rehabilitation' was implemented, drawing on combined medical, industrial and social expertise.[40] *Civilians into Soldiers* builds on this scholarship by exploring the role of medicine and healthcare before men went into active service. By examining regimes such as diet, rest and exercise, it highlights the healthy, fully functioning military body as a key site of medical intervention. It explores the relationships that were forged between civilian doctors and military experts and the contribution that medical science made to the prosecution of the war. However, it also highlights the opportunities that warfare afforded members of the medical community, who came to use the bodies of soldiers to develop their knowledge and skills. By focusing on the army as a site of health enhancement as well as treatment and repair, this work adds to wider debates about the relationship between war and medicine in the modern period.[41]

Civilians into Soldiers also considers the contribution of wider public health developments to military regimes of bodily management. As Dorothy Porter, Greta Jones, Stephen Constantine and Ina Zweiniger-Bargielowska have shown, the early twentieth century witnessed increasing state intervention in all citizens' bodies.[42] In light of the high numbers of men who were rejected as unfit for military service during the Boer and First World War recruitment campaigns, there were increasing fears of a declining national stock. These ideas focused particularly on the condition of the urban working classes. Thus, a whole set of practices and ideas emerged encompassing individual, family and industrial health. The 1904 report of the Interdepartmental Committee on Physical Deterioration had highlighted environmental factors as the cause of poor urban health and recommended a host of measures, including better nutrition, slum clearance and workplace reforms.[43] The welfare legislation passed by the Liberal governments of 1906 and 1916 expanded the remit of public health provision to include free school meals, medical inspections and compulsory physical education, as well as provisions for better housing, sanitation and food supply, maternity and child services and unemployment assistance. The creation of government agencies like the Ministry of Health in 1919 was also directed towards the increased management of bodies. Designed to consolidate all the medical and public health functions of central government, the ministry was responsible for the coordination and supervision of all local health services. From 1938 its role was greatly expanded as it took charge of the whole wartime Emergency Medical Services, including hospital care, ambulance provision, medical supplies and public health. The Ministry was also responsible for the provision of shelters, the administration of the civilian evacuation scheme and worked with the Military Recruitment Department of the Ministry of Labour and National Service, which medically examined men and women for service in the armed forces.[44]

It was not, however, just through government initiatives that bodies were reformed during this period. The physical culture movement, which had emerged in the late nineteenth century, thrived in Britain during the 1920s and 1930s. Zweiniger-Bargielowska explains that organisations like the National League of Physical Education and Improvement, the New Health Society, the Health and Strength League and the Sunlight League 'shared an ideological affinity' and advocated measures such as dietary and dress reform, regular exercise, personal cleanliness, sun-bathing,

Introduction

fresh air and a return to nature as a means of rescuing bodies from the effects of the urban environment.[45] Key to their ambitions was to educate people to invest in their own fitness and wellbeing in the belief in 'the cultivation of health as a moral and civic duty'.[46] The principles of physical culture transcended the boundaries of class and appealed to both men and women. It was embraced by the government in 1937 with the launch of a National Fitness Campaign to provide funding to local authorities and voluntary organisations for improved recreational facilities and the training of instructors.[47]

The increasing rationalisation of workers' bodies through new ideas of industrial organisation was also a key feature of early twentieth-century Britain. This was the era of 'scientific management', the system of labour organisation that analysed each worker's task and movements in minute detail in order to increase productivity.[48] Pioneered in America by Frederick Taylor, scientific management was disseminated in Britain by businessman Charles Bedaux.[49] According to the Marxist historian Harry Braverman, it was 'simply a means for management to achieve control of the actual mode of the performance of every labour activity'. Braverman claims that scientific management dehumanised workers, who were 'reduced almost to the level of labour in its animal form'.[50] Arthur McIvor and Vicky Long also suggest that occupational health and safety were greatly developed in Britain before and during the Second World War, as both the state and private managers came to realise that maintaining employee health was consistent with efficiency and profit maximisation. The First World War had hastened state intervention in industrial health through the creation of bodies like the Health of Munitions Workers Committee, which was set up to investigate fatigue, working hours and other issues that affected worker productivity. Although employers and trade unions had become the main proponents of workplace reforms during the interwar years, the outbreak of hostilities in 1939 meant that government officials once again came to intervene in the workplace in order to meet militaristic demands. Under the 1940 Factories (Medical and Welfare Services) Order, the Ministry of Labour and National Service was able to organise medical supervision, nursing and welfare and first-aid service for workers employed in munitions factories. Organisations like the Industrial Health Research Board, Industrial Health Advisory Committee and the Factory and Welfare Advisory Board advised on technical and scientific problems, investigated industrial diseases and made arrangements for the supervision of the general health and welfare of factory employees.[51]

Civilians into Soldiers links all of these wider developments surrounding civilian health to the wartime army. It shows that in an era when one-fifth of the working male population was recruited into the army's ranks, soldiers' bodies easily became a site around which wider public health debates could emerge.[52] It demonstrates how principles of physical culture permeated the military environment through dietary and exercise regimes. It also shows that military leaders often came to draw on ideas developed in the sphere of industry to improve manpower, efficiency and morale. As such, this study explores the place of the wartime army within the wider project of modernity in early twentieth-century Britain, in which the body became a key target of progress and reform.

Finally, gender is important to this study. While much has been written about the experiences of women and the constructions of femininity in Britain during the Second World War, literature on the relationship between war and masculinity has been largely confined to the conflict of 1914–18.[53] In her influential *Dismembering the Male*, Joanna Bourke examines the impact of the First World War on men's bodies and masculinity as well as male bonding in the armed services. She suggests that wartime experiences brought about a greater sharing of gender identities among men of different classes, ages and localities and 'a narrowing in the way men of all classes experienced their own corporeality'.[54] More recently, Ana Carden-Coyne has explored cultural representations of the reconstruction of disabled military bodies in Britain, the United States and Australia after the First World War. She focuses specifically on the influence of classicism – the classical imagery of human bodies – which merged with 'modern attitudes' and allowed people to come to terms with the human cost of the war.[55]

This book explores the relationship between the military body and masculinity between 1939 and 1945. Its aim is to understand how wider ideas about masculinity permeated the military environment. Graham Dawson suggests that the most extreme form of hegemonic masculinity has historically been found in the figure of the warrior-hero. He states that 'the soldier hero has proved to be one of the most durable and powerful forms of idealized masculinity within Western cultural traditions since the time of the ancient Greeks. Military virtues such as aggression, strength, courage and endurance have repeatedly been defined as the natural and inherent qualities of manhood, whose apogee is attainable only in battle.' Dawson argues that during the nineteenth century a Victorian version of the soldier-hero emerged with the rise of British imperialism and the

Introduction

new notions of the nation.⁵⁶ These ideas of martial masculinity were also extolled through boys' organisations like the Scouts and the Boys' Brigade, whose members wore uniforms and took part in military-style drilling and fighting games that emphasised qualities such as strength and discipline. These institutions played a key role in disseminating such Victorian martial-masculine ideals from their original location in the public schools to boys from the middle and working classes.⁵⁷ Groups like the Boys' Brigade and the Young Men's Christian Association (YMCA) promoted 'muscular Christianity', which added an overtly religious element to these models of martial-masculinity.⁵⁸ Zweiniger-Bargielowska suggests that this emphasis on the fit male body remained important into the twentieth century and throughout the interwar years influenced the physical culture movement, which prized physical fitness and muscular beauty as the epitome of imperial manliness and racial strength.⁵⁹ In her work, which focuses on the home front, Sonya Rose argues that during the Second World War these physical characteristics were combined with the virtues of restraint and chivalry to create what she has termed 'temperate masculinity'. This version of manliness applied only to men in uniform and emphasised the understated courage of the 'little man' which was constructed in opposition to the hyper-masculine Nazi.⁶⁰

Civilians into Soldiers examines how these cultural ideals played out in the context of the wartime army. I argue that regimes of military management often reflected wider social ideals and assumptions surrounding the male body in order to achieve their objectives. In a society that prized the strong, muscular physique, men's bodies were judged not just in terms of functionality but also in appearance. However, recent scholarship also proposes that masculinity is not homogeneous, but plural and diverse.⁶¹ Thus, in the military, different forms of masculinity can operate at different levels, such as between high-ranking officers and private recruits.⁶² Within the ranks, for example, conceptions of manliness have often centred on indulging in bodily pleasures and excesses such as getting drunk and having sex.⁶³ This book, therefore, explores the ways in which men at various levels of the military hierarchy fashioned their own gender identities and how these intersected with wider discourse. As such, this work highlights the importance of the body to various performances of masculinity among wartime recruits.

When all of these historiographies are drawn together they not only present a range of approaches to the material but also questions to be answered by it. *Civilians into Soldiers* tackles issues related to the objectives, the agendas, the aesthetics and the techniques of those in the British Army of the Second World War when they sought to control and transform male bodies. By doing so it will answer questions about the nature of military culture in modern Britain, the ways in which wider social practices shaped military life and the extent to which the British prepared for war in new or unusual ways.

Men's bodies: public and private representations

In order to examine official representations of the military body during the Second World War, *Civilians into Soldiers* draws on a wide range of government, military and medical records. Between 1939 and 1945 the departments of the War Office produced various army pamphlets, reports and correspondence, including those of the Army Medical Services, which give accounts of the activities of individual medical units in all theatres of operations. Proceedings of government departments such as the Ministry of Labour and National Service also show how issues pertaining to military health and fitness were dealt with, from the initial stages of recruitment to procedures for applying for disability pensions. Military and medical ideas about the body are also evident in professional journals of the day. These have not been read in order to trace an objective reality, but as instances where ideas have been produced within discourses implicit to the wartime military and medical project. As Bryan Turner points out, the 'medical gaze' allows dominant groups to impose moral judgements with an air of scientific knowledge.[64] Howard Waitzkin also states that within medical sources 'symptoms, signs and treatment take on an aura of scientific fact.'[65] In using these materials, this work is therefore careful to consider the wider objectives of the individuals and the organisations that produced them.

It is more difficult to recover the 'voices emanating from the bodies themselves'.[66] In order to do so this book draws upon a range of servicemen's personal testimonies in both oral and written form. These include various reports, letters and diaries from the Mass Observation project, the social research programme that recorded people's everyday lives in Britain between 1937 and 1948.[67] During the course of the Second World War several volunteer observers served in the army and reported on various aspects of day-to-day life in the forces. These men were also sent monthly 'directives' consisting of lists of questions to ask their fellow recruits and report back on.[68] Full-time observers who did not serve in the military also recorded the overheard comments of men in the forces as well as having conversations and conducting interviews directly with them. These included discussions of issues such as military medical examinations, sex life in the forces and men's responses to physical training. However, as Penny Summerfield suggests, Mass Observation does not provide a wide representation of everyday experience as the majority of volunteers were young members of the lower middle class, many of whom were socio-politically motivated.[69] Wanting to be part of the fight against fascism, they aimed to draw attention to social conditions at home and opposed the official neglect of ordinary people.[70]

In order to access a wider range of experiences, several published and unpublished autobiographies and written testimonies collected by the BBC WW2 People's War Archive between 2003 and 2006 are also included in this study. Written accounts like these provide great insight into the intricacies of military participation. Rachel Woodward and Neil Jennings suggest that military memoirs inform us about the 'the smaller details of human experience within broader historical narratives of armed service'.[71] The value of autobiographical accounts to studies of embodied experience is also highlighted by Sidone Smith and Julia Watson, who point out that 'Life narrative is a site of embodied knowledge because autobiographical narrators are embodied subjects. Life narrative inextricably links memory, subjectivity and the materiality of the body.'[72] It is, however, primarily through a selection of oral history interviews from the Imperial War Museums Sound Archive (IWM SA) that the bodily experiences of wartime recruits are explored within this work. These structured interviews were conducted mainly between the 1980s and the early twenty-first century and consist of whole war narratives, from enlistment to demobilisation. They include the responses of both officers and ordinary soldiers, from various parts of the United Kingdom, and from a range of social backgrounds and occupations. Their journeys from recruitment station to battlefield are the corporeal experience that forms the basis of this book.

Oral history has proved crucial as a means by which to access the experiences of those who have not been included in the official record, such as women, ethnic minorities and ordinary working people, and its value has been widely recognised by historians.[73] Jan Walmsley and Dorothy Atkinson suggest that 'oral history can be both "more history", adding to the stock of knowledge about historical events, and "anti-history", challenging conventional perceptions.'[74] Importantly for this work, Nigel de Lee emphasises that 'oral history can address some obvious, important and obscure questions about war: questions which are not susceptible to exact measurement, calculation or cold logic.'[75] Indeed, oral histories have provided new understandings about the impact of war on social change, as is shown in the work of Joanna Bourke and Penny Summerfield.[76] Oral history is also particularly valuable to studies of medicine, health and welfare, all of which are key themes in this book. Paul Thompson notes that 'oral history can delve into the hidden world of the institution, the clinic or the hospital, revealing the daily experience of routines and treatments as told by the subjects, clients or patients at the receiving end of services.'[77] It is therefore by drawing on their oral testimonies that *Civilians into Soldiers* seeks to understand how soldiers experienced medical care, how they perceived medical and scientific technologies and how they reacted to the army's efforts to manage their health and wellbeing.

The use of personal recollections does, nevertheless, pose several problems, which transgress both the written and spoken material. Firstly, the reliability of memory must be considered when examining the testimonies that were not produced during the war but are works of subsequent reflection. Joanna Bourke states that 'men who were there claim a higher knowledge than other commentators, yet in the heat of battle experiences were often confused, indeterminate and unarticulated.'[78] The body, as the focus of inquiry, can also be problematic for a study based on soldiers' first-hand accounts. This is due to the assumption that the body is largely 'absent' from consciousness.[79] Drew Leder suggests that individuals are not aware of their bodies unless they are in 'dysfunctional states' such as pain or suffering. He asserts that, 'while in one sense the body is the most abiding and inescapable presence in our lives, it is also characterised by absence. That is one's own body is rarely the thematic object of experience. When reading a book or lost in thought, my own bodily state may be the furthest thing from my awareness...the body, as a ground of experience...tends to recede from direct experience.'[80]

Yet, while the body can be challenging to explore within personal testimonies, this does not mean that it simply cannot be found. Nettleton and Watson argue that, although the body is largely 'taken for granted' in everyday life, people do become aware of their bodies in a range of circumstances and not just in moments of breakdown or stress. For instance, individuals are often aware of their own physical transformations, such as when growing stronger or fitter.[81] As this work will show, these were key features of army life that men discussed frequently in their diaries, memoirs and interviews. Moreover, Rosalind

Introduction

Gill, Karen Henwood and Carl McLean suggest that rather than talking about flesh and blood per se, men talk about body-related behaviour as a way of making sense of their embodied experience. These scholars argue that in modern society young men must work on their bodies while simultaneously disavowing any (inappropriate) interest in their own appearance. By talking about bodily related practices, such as sport, working-out and tattooing, rather than skin or muscle, men actively engage in constructing and policing appropriate masculine behaviours. By looking for particular bodily behaviours and contexts, the researcher can therefore see how the embodied self is expressed.[82]

Indeed, as the personal testimonies used in this work were not the product of deliberate interrogation, their inclusion serves to highlight the centrality of the body in recollections of war. The approach taken is not to treat these testimonies as eye-witness accounts of particular events or campaigns, but rather to analyse how soldiers have articulated and constructed their narratives of military service.[83] *Civilians into Soldiers* explores questions such as: How did men feel about being accepted or rejected for the army? What were their thoughts about going into battle for the first time? How did they feel when confronted with the enemy and how did they react to seeing dead comrades? This work is also careful to consider soldiers' testimonies within the wider social and cultural contexts in which certain behaviours took place.[84] Kathleen Canning suggests that it is the job of the historian to 'untangle the relationships between discourses and experiences by exploring the ways in which subjects mediated or transformed discourses in specific historical settings'.[85] This is not always straightforward because competing ideas all affect individual attitudes and beliefs. Different social groups, such as men, women and people of different social classes, may also be influenced differently by wider cultural values and ideals. This allows individuals some opportunity for 'selection' or 'rejection' of the available discursive understandings of themselves and their societies.[86] Thus, when analysing their accounts of military participation, *Civilians into Soldiers* considers how men were influenced by dominant military discourse such as heroism, sacrifice, duty and comradeship, as well as wider cultural assumptions about gender, sexuality, nationhood and citizenship. It examines how men have remembered the physical aspects of army life and how they have expressed their acceptance or rejection of particular ideals and norms. In doing so, this work explores how the embodied experience of military service, from initial training to armed combat, has been represented by soldiers themselves.

Chapter outline

Both institutional records and personal testimonies consistently place the body at the heart of narratives about military experience during the Second World War. As such, this book is constructed around the sequence by which the body was classified, controlled and then converted for combat, before being sent into action.

Chapter 1, 'Examination', explores the army's physical selection process, the first point in the transformation from civilian to soldier. It begins by outlining the methods that were used by examining doctors and the physical qualities that were considered desirable for recruitment. These suggest that the medical exam was not an objective assessment of the male body but was shaped by a range of practical concerns and political agendas. The chapter then draws on the experiences of men who entered this sorting system. These suggest that examination was often a moment of contest and negotiation between the individual and the wartime state, as men attempted to manage their encounters with medical examiners, either to avoid or to secure enlistment.

Rachel Woodward suggests that 'the training areas and barrack rooms produce the soldier's body.'[87] In Chapter 2 the processes by which civilian bodies were trained and adapted for military use are explored. I suggest that army training was a two-stage process of control and transformation. First, army leaders imposed a range of interventions upon the recruit's body in order to subject him to the authority of the regime. Instructors then proceeded to transform his body into an effective military machine by making it fit, ordered and productive. This was achieved through a strict regime of physical exercise, field exercises, team sports and military drill. However, even within the confines of the training camp or barracks, there was great potential for agency on the part of the individual soldier. While some enjoyed growing fitter, eating more and developing new skills, others chose to engage in what were deemed dangerous or inappropriate bodily behaviours. As such, this chapter considers the dynamics of compliance, resistance and participation in modern regimes of the state through corporeal transformation.

The medical and scientific experiments conducted on army personnel are the focus of Chapter 3. The wartime human research programme included trials of therapeutic drugs, synthetic stimulants and exposure to chemical agents. All of these were planned and carried out by government and military agencies like the Medical Research Council (MRC) and the Chemical Defence Experimental Establishment at Porton Down.

Introduction

The experiments on army personnel provide glimpses into the mind-set and decisions made by researchers regarding the types of bodies that were considered most useful and the levels of risk to which they were to be exposed. They also highlight the objectification of the military body as it became a scientific specimen with which to further the quest for manpower efficiency and wider medical knowledge. This chapter reveals, however, that as a site of experimentation the soldier's body was not simply subjected to the needs of science. By exploring the ways in which servicemen encountered medical and scientific experiments, it challenges much of the existing historiography, which highlights human experiments as key instances of military and state coercion. Rather, I suggest that there were a wide range of reasons that influenced how and why men came to take part in human trials and that they had clear ethical expectations for their own bodies when they did so. As such, this chapter highlights the active role that soldiers played in shaping wartime research as they engaged, and indeed withdrew, their bodies in the demands of experimental science.

In Chapter 4, 'Active service', the continued management of the soldier's body in the field of operations is explored. Here I examine the main ideals and perceptions of men's bodies that operated in theatres of war. Specifically, I demonstrate that the authorities overseas were influenced by older colonial ideas about physical health and fitness. Military officials emphasised the vulnerability of British troops to changed climates and landscapes, as well as to the dangerous 'other' bodies that the men encountered. An array of techniques and practices thus emerged in order to protect British personnel. As will be seen, however, the ways in which men's bodies were maintained in the field of operations represented a marked shift from the strict control that had existed in the army camp or barracks. In an environment where soldiers were not always under the immediate command of superiors, the army came to rely more on self-regulation through more subtle strategies. This chapter thus explores the army's wartime campaign of health education and propaganda, which encouraged men to look after their own health and fitness and taught them the skills by which to do so. Soldiers' testimonies show, however, that the army could not always achieve its objectives. While some men chose to follow the advice they were given, many others preferred to ignore it, deciding instead to get drunk, have sex or not to use treatments or prophylactics. By exploring these sorts of behaviours, and the motivations behind them, this chapter highlights the changing nature of military life in active service and the new expectations that men had for their bodies once released into theatres abroad.

The final chapter considers the place of the body in official and individual responses to fear, wounding and death. It examines how the army tried to control men's emotions through bodily channels and the processing of wounded and dead bodies through complex medical administrative instructions and burial regulations. The chapter then goes on to explore how individual soldiers coped with moments of problematic bodily performance: how they dealt with the breakdown or failure of their own bodies and their responses to the sight of death of others. These experiences, I suggest, could have a profound impact on fixed notions of the self and rendered many soldiers unable to continue with their military tasks. This chapter therefore highlights the individualism of the military body on the front line, despite all of the army's efforts to discipline and control it. Finally, the chapter examines the commodification of dead and wounded bodies through the military pensions system. While existing work has focused on the care, treatment and rehabilitation of wounded and disabled bodies during the war, the compensation issue has not yet been the focus of historical inquiry.[88] I argue that that the military pensions system is important because it also highlights the continued monitoring and surveillance of men's bodies after military service was over. Although it was no longer of any use to the army, the body continued to be of interest to the state. It was routinely examined, classified and categorised in order to safeguard against financial losses.

By exploring these various contexts, this book thus highlights the centrality of the body to the transition from civilian to soldier between 1939 and 1945. It examines the aims, ambitions and aspirations of government officials, military and civilian doctors, recruiters, instructors and soldiers themselves, all of whom played a crucial role in the body's transformation. In doing so, *Civilians into Soldiers* draws upon and extends a number of historiographies, including historical engagements with the place of the body in modernity, the role of medicine in the early twentieth century, the history of the British Army and the social and cultural history of the Second World War.

Notes

1 Imperial War Museums Sound Archive, London (hereafter IWM SA), 23195, Roy Bolton, reel 2.
2 The National Archives, Kew (hereafter TNA) WO277/12, Army and A.T.S., 1939–46, appendix C, p. 80.
3 M. Foucault, *Discipline and Punish: The Birth of the Prison* (London:

Introduction

Penguin, 1977).
4   Foucault, *Discipline and Punish*, p. 135.
5   M. Foucault, 'Technologies of the self', in L.H. Martin and H. Gutman (eds.), *Technologies of the Self: A Seminar with Michel Foucault* (Amherst, MA: University of Massachusetts Press, 1988), p. 18.
6   M. Foucault, 'Governmentality', in G. Burchell, C. Gordon and P. Miller (eds.), *The Foucault Effect: Studies in Governmentality'* (Chicago: University of Chicago Press, 1991), pp. 102–3. See also, J. Coveney, 'The government and ethics of health promotion: the importance of Michel Foucault', *Health Education Research, Theory and Practice* 13:3 (1998), 461.
7   C. Shilling, *The Body and Social Theory* (London: Sage, 2nd edn, 1993), pp. 10–11.
8   R.W. Connell, *Masculinities* (Cambridge: Polity, 2nd edn, 1995), p. 51.
9   Connell, *Masculinities*, p. 51.
10 B. Turner, *Regulating Bodies: Essays in Medical Sociology* (London: Routledge, 1992), p. 16.
11 Turner, *Regulating Bodies*, p. 21.
12 See, for example, S. Newman, *Embodied History: The Lives of the Poor in Early Philadelphia* (Philadelphia: University of Pennsylvania Press, 2003); J. Bullington, 'Body and self: a phenomenological study on the ageing body and identity', *Medical Humanities* 32 (2006), 25–31; R. Gill, K. Henwood and C. McLean, 'Body projects and the regulation of normative masculinity', *Body and Society* 11:1 (2005), 37–62.
13 P. Bourdieu, *Distinction: A Social Critique of the Judgment of Taste* (London: Routledge, 1984), pp. 2–6, 217–18.
14 E. Goffman, *Behaviour in Public Places: Notes on the Social Organization of Gatherings* (New York: Free Press, 1963), pp. 33–5.
15 D. Leder (ed.) *The Body in Medical Thought and Practice* (London: Kluwer Academic, 1992), p. 25.
16 M. Merleau-Ponty, *The Phenomenology of Perception*, trans. Colin Smith (London: Routledge, 2nd edn, 2002), p. 210.
17 N. Crossley, 'Body-subject/body-power: agency, inscription and control in Foucault and Merleau-Ponty', *Body and Society* 2:2 (1996), 101.
18 A. Frank, 'For a sociology of the body: an analytical review', in M. Featherstone, M. Hepworth and B. Turner (eds.), *The Body: Social Processes and Cultural Theory* (London: Sage, 1991), p. 58.
19 Connell, *Masculinities*, p. 56.
20 J. Hockey, 'Head down, bergen on, mind in neutral: the infantry body', *Journal of Political and Military Sociology* 30:1 (2002), 148–71.
21 Hockey, 'Head down, bergen on', 152.
22 P. Higate, 'The body resists: everyday clerking and unmilitary practice', in S. Nettleton and J. Watson (eds.), *The Body in Everyday Life* (London: Routledge, 1998), pp. 180–98.

23 Higate, 'The body resists', p. 189.
24 J.A. Hollander and R.L. Einwohner, 'Conceptualizing resistance', *Sociological Forum* 19 (2004), 539–42.
25 J.C. Scott, *Domination and the Arts of Resistance: The Hidden Transcript* (New Haven, CT: Yale University Press, 1990), pp. 1–10.
26 Scott, *Domination and the Arts of Resistance*, p. 10.
27 N. Armstrong and E. Murphy, 'Conceptualising resistance', *Health* (2011), 5.
28 Hollander and Einwohner, 'Conceptualizing resistance', 543.
29 Crossley, 'Body-subject/body-power', 109–10.
30 D. Leder, *The Absent Body* (Chicago: University of Chicago Press, 1990), p. 84.
31 Recent publications include: R. Doherty, *Ubique: The Royal Artillery in the Second World War* (Stroud: History Press, 2008); C. Whiting and E. Taylor, *The Fighting Tykes: An Informal History of the Yorkshire Regiments in the Second World War* (Barnsley: Pen and Sword Military, 2008); J. Devine, *Forgotten Voices of Dunkirk* (London: Ebury Press, 2010); A. Beevor, *D-Day: The Battle for Normandy* (London: Penguin, 2010); F. McLynn, *The Burma Campaign: Disaster into Triumph, 1942–45* (London: Bodley Head, 2010); P. Delaforce, *Churchill's Desert Rats in West Africa and Italy* (Barnsley: Pen and Sword Military, 2009).
32 J.B. Davis (ed.), *Grand Campaigns of World War II* (Leicester: Silverdale Books, 2002); D. Fraser, *And we Shall Shock them: The British Army in the Second World War* (London: Cassell Military, 1983); T. Harrison-Place, *Military Training in the British Army, 1940–1944: From Dunkirk to D-Day* (London: Frank Cass, 2000); J. Keegan, *Churchill's Generals* (London: Warner, 1992); J. Keegan, *The Second World War* (London: Pimlico, 1997).
33 J. Crang, *The British Army and the People's War* (Manchester: Manchester University Press, 2000), p. 2.
34 J. Crang, 'Square pegs in round holes: other rank selection in the British Army, 1939–45', *Journal of the Society for Army Historical Research* 77 (1999), 293–8.
35 Crang, *The British Army and the People's War*, pp. 64–5.
36 Crang, *The British Army and the People's War*, pp. 139–42.
37 D. French, *Raising Churchill's Army: The British Army and the War against Germany, 1919–1945* (Oxford: Oxford University Press, 2000), p. 11.
38 French, *Raising Churchill's Army*, pp. 285–6.
39 M. Harrison, *Medicine and Victory: British Military Medicine in the Second World War* (Oxford: Oxford University Press, 2004), p. 2. For a more detailed discussion of the concept of rationalization see H.H. Gerth and C. Wright Mills (eds.), *From Max Weber: Essays in Sociology* (London: Routledge, 1948), p. 50.
40 J. Anderson, *War, Disability and Rehabilitation in Britain: 'Soul of a Nation'* (Manchester: Manchester University Press, 2011), pp. 4, 44.

Introduction

41 R. Cooter, 'Medicine and the goodness of war', *Canadian Bulletin of Medical History* 12 (1990), 147–59.
42 G. Jones, *Social Hygiene in Twentieth-Century Britain* (London: Croom Helm, 1986), p. 29; S. Constantine, *Social Conditions in Britain, 1918–1939* (London: Methuen, 1983), p. 35; D. Porter, *Health, Civilization and the State: A History of Public Health from Ancient to Modern Times* (London: Routledge, 1999), pp. 165–7; I. Zweiniger-Bargielowska, *Managing the Body: Beauty, Health and Fitness in Britain, 1880–1939* (Oxford: Oxford University Press, 2010).
43 Zweiniger-Bargielowska, *Managing the Body*, pp. 71–6.
44 See A. S. MacNalty, *The Civilian Health and Medical Services, Volume 1: The Ministry of Health Services; Other Civilian Health and Medical Services* (London: HMSO, 1953).
45 Zweiniger-Bargielowska, *Managing the Body*, pp. 161–92.
46 Zweiniger-Bargielowska, *Managing the Body*, pp. 161–2.
47 I. Zweiniger-Bargielowska, 'Building a British superman: physical culture in interwar Britain', *Journal of Contemporary History* 41:4 (2006), 607–8.
48 A. Rabinbach, *Human Motor: Energy, Fatigue and the Origins of Modernity* (New York: Basic Books, 1990), pp. 238–44.
49 F.W. Taylor, *The Principles of Scientific Management* (New York and London: Harper, 1911); S. Kreis, 'The diffusion of scientific management: the Bedaux Company in America and Britain, 1926–1945', in S. Kreis (ed.), *A Mental Revolution: Scientific Management since Taylor* (Columbus: Ohio State University Press, 1992), pp. 156–74.
50 H. Braverman, *Labor and Monopoly Capital: The Degradation of Work in the Twentieth Century* (New York: Monthly Review Press, 1974), p. 113.
51 V. Long, *The Rise and Fall of the Healthy Factory: The Politics of Industrial Health in Britain, 1914–1960* (London: Palgrave, 2010), pp. 16–48; A.J. McIvor, *A History of Work in Britain, 1880–1950* (London: Palgrave, 2001), pp. 111–47.
52 *Strength and Casualties of the Armed Forces and Auxiliary Services of the United Kingdom 1939-1945* (1946, Cmd. 6832), p. 4; Ministry of Labour and National Service, *Report for the Years 1939-1946* (1946–7, Cmd. 7225), p. 2.
53 H.L. Smith, 'The womanpower problem in Britain during the Second World War', *Historical Journal* 27:4 (1984), 925–45; P. Summerfield, *Women Workers in the Second World War: Production and Patriarchy in Conflict* (London: Croom Helm, 1984); P. Summerfield, *Reconstructing Women's Wartime Lives: Discourse and Subjectivity in Oral Histories of the Second World War* (Manchester: Manchester University Press, 1998); A. Lant, *Blackout: Reinventing Women for Wartime British Cinema* (Princeton: Princeton University Press, 1991); C. Gledhill and G. Swanson, *Nationalising Femininity: Culture, Sexuality, and British Cinema in the*

Civilians into soldiers

Second World War (Manchester: Manchester University Press, 1996). More recently, studies have emerged on masculinity in the Home Guard between 1939 and 1945, see P. Summerfield and A. Carden-Coyne, *Contesting Home Defence: Men, Women and the Home Guard in the Second World War* (Manchester: Manchester University Press, 2007), and in the Royal Air Force, see M. Francis, *The Flyer: British Culture and the Royal Air Force, 1939–1945* (Oxford: Oxford University Press, 2008).

54 J. Bourke, *Dismembering the Male: Men's Bodies, Britain and the Great War* (London: Reaktion, 1996), pp. 30, 251.
55 A. Carden-Coyne, *Reconstructing the Body, Classicism, Modernism and the First World War* (Oxford: Oxford University Press, 2009), p. 2.
56 G. Dawson, *Soldier Heroes: British Adventure and the Imagining of Masculinities* (London: Routledge, 1994), p. 1.
57 J. Weeks, *Sex, Politics and Society: The Regulation of Society Since 1800* (London: Longman, 1981), p. 40; J.A. Mangan and J. Walvin, 'Introduction', in J.A. Mangan and J. Walvin (eds.), *Manliness and Morality: Middle-Class Masculinity in Britain and America* (New York: St Martin's Press, 1987), p. 5.
58 For a more indepth discussion of 'muscular Christianity' see J. Springhall, 'Building character in the British boy: the attempt to extend Christian manliness to working class adolescents', in Mangan and Walvin (eds.), *Manliness and Morality*, pp. 52–74.
59 Zweiniger-Bargielowska, 'Building a British superman', 595–7.
60 S.O. Rose, *Which People's War? National Identity and Citizenship in Wartime Britain, 1939–1945* (Oxford: Oxford University Press, 2003), pp. 159–68.
61 R.W. Connell, *The Men and the Boys* (Cambridge: Polity, 2000), p. 10.
62 P.R. Higate, 'Introduction: putting men and the military on the agenda', in P.R. Higate (ed.), *Military Masculinities: Identity and the State* (Westport, CT: Praeger, 2003), p. xvii; R. Woodward and T. Winter, *Sexing the Soldier: The Politics of Gender and the Contemporary British Army* (London: Routledge, 2007), p. 62.
63 See for example, J. Hockey, 'No more heroes: masculinity in the infantry', in Higate (ed.), *Military Masculinities*, pp. 22–3; D. Morgan, 'Theater of war: combat, the military and masculinities', in H. Brod and M. Kaufman (eds.), *Theorising Masculinities* (London: Sage, 1994), p. 167.
64 Turner, *Regulating Bodies*, p. 18.
65 H. Waitzkin, 'A critical theory of medical discourse: ideology, social control and the processing of social context in medical encounters', *Journal of Health and Social Behaviour* 30:2 (1989), 224.
66 S. Nettleton and J. Watson, 'The body in everyday life: an introduction', in Nettleton and Watson (eds.), *The Body in Everyday Life*, p. 12
67 L. Noakes, *War and the British: Gender, Memory and National Identity* (London: I.B. Tauris, 1998), p. 75.

Introduction

68 P. Summerfield, 'Mass Observation: social research or social movement?', *Journal of Contemporary History* 20:3 (1985), 441.
69 Summerfield, 'Mass Observation', 442.
70 Summerfield, 'Mass Observation', 445.
71 R. Woodward and N. Jennings, 'Soldiers' bodies and the contemporary British military memoir', in K. McSorely (ed.), *War and the Body* (London: Routledge, 2013), p. 152.
72 S. Smith and J. Watson, *Reading Autobiography* (Minneapolis: University of Minnesota Press, 2001), p. 37.
73 See, for example, S.H. Armitage and S. Berger Gluck, 'Reflection on women's oral history: an exchange', in R. Perks and A. Thomson (eds.), *The Oral History Reader* (London: Routledge, 2nd edn, 2006), pp. 73–82; A. Portelli, *They Say in Harlan County: An Oral History* (Oxford: Oxford University Press, 2010).
74 J. Walmsley and D. Atkinson, 'Oral history and the history of learning disability', in J. Bornat, R. Perks, P. Thompson and J. Walmsley (eds.), *Oral History, Health and Welfare* (London: Routledge, 2000), p. 180.
75 N. de Lee, 'Oral history and British soldiers' experience', in P. Addison and A. Calder, *Time to Kill: The Soldier's Experience of War in the West* (London: Pimlico, 1997), p. 365.
76 Bourke, *Dismembering the Male*; Summerfield, *Women Workers in the Second World War*.
77 P. Thompson, 'Introduction', in Bornat, Perks, Thompson and Walmsley (eds.), *Oral History, Health and Welfare*, p. 4.
78 J. Bourke, *An Intimate History of Killing: Face-to-Face Killing in Twentieth-Century Warfare* (London: Granta, 1999), p. 9.
79 Leder, *The Absent Body*, p. 1.
80 Leder, *The Absent Body*, p. 1.
81 Nettleton and Watson, 'The body in everyday life: an introduction', p. 10.
82 Gill, Henwood and McLean, 'Body projects and the regulation of normative masculinity'.
83 R. Perks and A. Thomson, 'Critical developments: introduction', in Perks and Thomson (eds.), *The Oral History Reader*, p. 2.
84 Perks and Thomson, 'Critical developments: introduction'.
85 K. Canning, 'Feminist history after the linguistic turn: historicizing discourse and experience', *Signs: Journal of Women in Culture and Society* 19:2 (1994), 373–74.
86 Summerfield, *Reconstructing Women's Wartime Lives*, pp. 12–15.
87 R. Woodward, 'Locating military masculinities: space, place and the formation of gender identity in the British Army', in Higate (ed.), *Military Masculinities*, p. 51.
88 Anderson, *War, Disability and Rehabilitation*.

1

Examination

Between 1939 and 1945 the bodies of over six and a half million men in Britain were medically examined in order to assess their suitability for military service. These included almost one and a half million volunteers and over five million men who were called up.[1] Under the Military Training Act of May 1939 all men aged twenty and twenty-one years old were required to undergo six months' training in the armed forces. Upon the outbreak of war in September this was superseded by the National Service (Armed Forces) Act, which imposed liability for military service on all males aged eighteen to forty-one, unless they were working in a reserved occupation or registered as conscientious objectors.[2] Each man was required to register at one of the 1,225 recruiting offices set up by the Ministry of Labour and National Service. There he was required to give his date of birth, occupation and employer, and if he wished, to express a preference for the service he wished to join. After the initial sorting of registration, he was summoned to a local recruiting centre, where he was medically examined and interviewed by a recruiting officer who decided whether he would be accepted for a particular service, after which he was posted.[3]

This chapter explores the methods and standards adopted by examiners in order to classify men's bodies and what these classifications reveal about how the male body was constructed as useful in wartime Britain. As will be seen, this apparently objective scientific encounter intensified state monitoring and surveillance of the body during war, as a range of social, political and practical agendas affected the ways in which men were measured, judged and graded. War also heightened wider public interest in the body. Government officials and medical experts debated the techniques and criteria adopted, and what the data gathered from examinations revealed about the nation's health.

Examination

A further purpose of this chapter is to explore what the medical exam meant for those men who underwent it. It looks at the ways in which individuals responded to the classifications applied to their bodies and how they tried to manipulate their encounters with medical examiners in order to secure or avoid enlistment. In his study of the military medical exam in the First World War, David Silbey has argued that this was not a 'straightforward filter' but 'a highly negotiated and contested gateway; one manipulated by both sides'. Doctors, patriotic and overwhelmed by numbers, aimed to pass as many men as they could, while enlistees, who were eager to serve but below the standards that the army had set, wished to get past the exam in any way they could.[4] The chapter also considers the medical exam of the Second World War as a site of bargaining and resistance in which the body was a key element. Rather than an impartial assessment of objective physiological data, I suggest that the exam was often a moment in which science was overturned in order to fulfil the needs of both the civilian and the wartime state.

Classifying the body

During the war, the task of classifying men's bodies in order to determine their fitness for service lay with civilian medical boards. Established by the Ministry of Labour and National Service, each consisted of a chairman and four other members. These were medical practitioners, usually over fifty years of age, who were selected by regional medical officers and local war committees.[5] As a general rule, at the start of the war, the boards examined thirty men in every two-and-a-half-hour session and held two sessions each day. Examinees stripped to their underwear and moved from doctor to doctor, each of whom was responsible for inspecting different bodily parts and functions. As one prospective recruit later recalled, 'they went through you like a dose of salts'.[6] Examiner No. 1 investigated mental condition, nervous stability and tested vision, before enquiring into the man's previous health, any hospital treatment and whether he had ever received a disability pension. Examiner No. 2 took weight, height and chest measurements, noting down hair and eye colour and any external marks such as scars and tattoos. He also looked at the lungs, recorded the pulse and assessed the eyes, nose, throat and teeth. Examiner No. 3 checked the heart and the pulse rate, paying attention to physical development and investigating deformities. Finally, Examiner No. 4 looked at the ears, tested the hearing and urine, checked the abdomen and re-examined the heart.[7] At the end, the whole board determined the man's final medical grade. In cases of disagreement or doubt it was left to the chairman to make a final decision. The examinee would then be placed into one of the following four medical categories:

Grade I – Subject only to such minor disabilities as can be remedied or adequately compensated by artificial means, attain the full normal standard of health and strength, and are capable of enduring physical exertion suitable to their age.

Grade II – Those who, while suffering from disabilities disqualifying them for grade I do not suffer from progressive organic disease, have fair hearing and vision, are of moderate muscular development and are able to undergo a considerable amount of physical exertion not involving severe strain. Where a man has been placed in this grade solely on account of either defects of visual acuity or deformities of lower extremities, or both, in accordance with the instructions in the appropriate paragraphs of this code, this will be signified by the letter (a) followed by the words vision or feet in brackets, e.g., grade II (a) (vision) or grade II (a) (feet).

Grade III – Those who present such marked disabilities or evidence of past disease that they are not fit for the amount of exercise required for grade II.

Grade IV – Those who suffer from organic progressive disease or are for other reasons permanently incapable of the kind of degree of exertion required for grade III. These men are unfit for any form of service.[8]

Capital letters A to D were also used to signify whether a man was fit for service at home or abroad:

A – Fit for service at home or abroad
B – Unfit for service abroad but fit for base or garrison service at home and abroad
C – Fit for home service only
D – Unfit for any form of service[9]

This system of classification was based on a code of instructions created during the First World War. After the introduction of conscription in 1916, Ministry of Labour medical boards had been assembled in order to examine men's bodies and categorise them into four medical grades ranging from I to IV. In 1939 there was, however, one key difference: the inclusion of men with defects of vision or feet in Grade II, and signified by the letter (a). During the First World War, these men had been categorised as Grade III.[10] Discussing this change in parliament in June 1939, Minister of Labour Ernest Brown explained that these new classifications took into greater account an individual's specific military

Examination

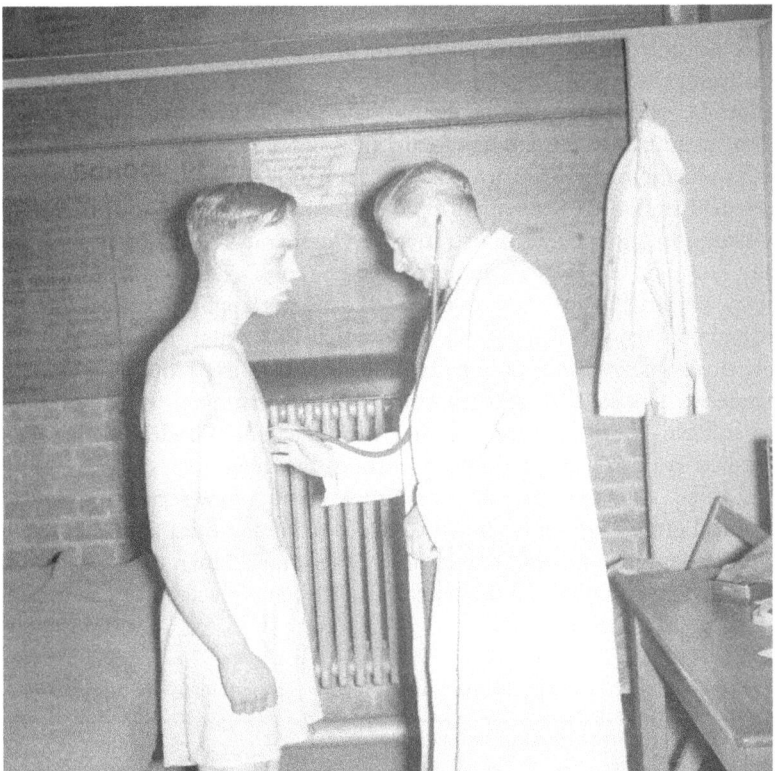

Figure 1 A prospective recruit undergoes a medical examination at the drill hall in the Old Dukes Road, Euston

abilities. A man with only one eye could, for instance, qualify as Grade II provided that the eye was 'free from disease and reaches a specified degree of visual activity'. The rationale being, in Brown's words, that 'I knew a Devonshire soldier in pre-war days who had only one eye and was an expert Bisley shot' and that 'Nelson had only one'.[11]

This emphasis on evaluating bodies according to military capacity quickly intensified as recruiting got underway and the government sought to make effective use of all available men. From the outset, one of the most pressing problems that the authorities faced was the selection of soldiers for a wide variety of roles and occupations within a modern technological army, each of which required different skills and capabilities.[12] Yet, unlike regimental military examiners in peacetime, the function of the civilian boards was not to assess men for entry into specific units but only to classify them into one of the four general medical grades. The *Lancet* noted in July 1939 that 'the army examination aims at determining whether a candidate is fit for the unit he applies to join; the militia examination whether he is fit to be a soldier at all.'[13] In the war's official medical histories, F.A.E. Crew also explained that 'the same physical standards were applicable to all units in a fighting formation, artillery, infantry or administrative services, and also to every man in any one unit, no matter whether rifleman, cook, clerk or batman.'[14] As such, within the first few weeks of the exam's application there were complaints from commanding officers that their intakes included men physically incapable of performing the specific duties required of them. Some men posted to field-training units, for example, were found to be suffering from hernias. They had been placed in Grade I; hence this implied no error in grading or posting, but nevertheless, these men had to be invalided out.[15] So, in February 1940 a new series of military medical grades were introduced which took into account an individual's ability to march, to shoot and to drive. This allowed men to be posted to appropriate army roles and occupations, by being placed into one of ten categories (Table 1).

Table 1 Medical classifications of soldiers by categories, February 1940

Army category	Army standard as regards physique and capabilities	Locality in which men may normally be employed
A1	See to shoot or drive Can undergo severe strain Without defects of locomotion With only minor (remediable) disabilities	Any area in a theatre of war
A2	See to shoot or drive Can undergo severe strain With slight defects of locomotion With only minor (remediable) disabilities	Any area in a theatre of war
B1	See to shoot or drive Can undergo considerable exertion not involving severe strain Without defects of locomotion With moderate degree of disabilities	Lines of communication (L of C), base, or garrison service at home or abroad

Examination

Army category	Army standard as regards physique and capabilities	Locality in which men may normally be employed
B2	See to shoot or drive Can undergo considerable exertion not involving severe strain Without defects of locomotion With moderate degree of disabilities	L of C, base, garrison service at home or abroad
B3	See for ordinary purposes Can undergo severe strain Without defects of locomotion With only minor (remediable) disabilities	L of C, base, or garrison service at home or abroad
B4	See for ordinary purposes Can undergo severe strain With slight defects of locomotion With only minor (remediable) disabilities	L of C, base, or garrison service at home or abroad
B5	See for ordinary purposes Can undergo considerable exertion not involving severe strain With or without defects of locomotion With moderate degree of disabilities	L of C, base, or garrison service at home or abroad
C	See for ordinary purposes Unfitted for considerable exertion With marked physical disabilities or evidence of past disease	Home service only
D	Temporarily unfit	
E	Permanently unfit	

Source: TNA WO293/25, Army Council Instruction (ACI) 184 of 1940, Instructions for the medical classification of soldiers by categories, 29 February 1940.

These categories were repeatedly expanded throughout the war as the numbers and quality of men available for military service declined.[16] In 1944, for instance, categories NT, 'non-tropical', and HS, 'home service', were introduced in order to denote the type of environment for which a man was physiologically suited, rather than distinguishing only between service at home and abroad. This meant that individuals with disabilities that were aggravated by hot climates could be sent to Europe, where the weather would not affect their physical performance, instead of having to remain in Britain. As a result of additions like these, by 1945 the medical grading system consisted of no fewer than 92 sub-categories.[17] Thus, physical classification was not fixed, but was constantly open to reinterpretation, according to the demands of war. In the quest for manpower economy the authorities placed less emphasis on the whole body and focused more on its respective components in order to determine what sorts of jobs men were suited to.

This conception of bodily usefulness based on occupational capacity was not unique to the military but reflected wider processes of rationalisation occurring in the civilian workforce. During the interwar years a range of government-sponsored industrial health research programmes had focused specifically on the psychological and physiological capabilities of employees in order to deploy labour most effectively. Anson Rabinach suggests that managers and trade union leaders looked increasingly to experts who could establish 'rational' physiological and psychological standards for selection in order to achieve greater conservation of 'human economy'.[18] Studies conducted in 1922 by the Industrial Fatigue Research Board had, for instance, illustrated 'the widely different qualities of the individual that may have to be taken into account in determining whether and how far he is suited for a particular type of work'.[19] These included an investigation into the physiological capacities of printing apprentices, a study of the physical strength required for different occupations and a study of hand measurements in sweet-factory workers.[20] Indeed, Jeremy Crang and David French have shown that during the war direct links were established between this kind of industrial health research and wider military selection. Intelligence tests introduced into the army were developed by members of the MRC's Industrial Health Research Board.[21] In 1941 a dedicated body, the Directorate for the Selection of Personnel, was also set up to devise more accurate selection procedures, with many of its members drawn from the National Institute of Industrial Psychology.[22] The latter had been created in 1921 to promote the application of psychology and physiology within industry and commerce. It was led by Charles Myers, a specialist in shell shock during the First World War.[23] Similar comparisons can also be made with physical classification. In 1940 seven new vision standards were added to the medical exam, which were based on the minimum standard required by the London Passenger Transport Board.[24] These ensured sufficient eyesight for driving purposes and for shooting up to a distance of 200 yards, again helping the army to assign individuals to particular military roles.[25] Through the medical examination process, the military body and that of the civilian worker therefore became more aligned during the war, as the authorities sought new ways of matching man and occupation.

It was not only the classifications applied to men's bodies that were open to reinterpretation during the war. Corinna Peniston-Bird suggests that, with the continuation of hostilities and as the needs of manpower intensified, the authorities also had to expand the pool from which recruits could be drawn.[26] One key way in which the state did this was by changing the age requirements for enlistment. From the outset of the war government officials had considered younger men as the most physiologically suited for military service. The Standing Committee on National Expenditure reported in the *British Medical Journal* in 1941, 'While the older man of military age had the advantage of greater experience, this is to a certain extent offset by relatively poor hand-and-eye co-ordination. Young men have proved of special value owing to their quickness of thought and action and also their greater physical fitness... Men of 18 and 19 are the most physically fit and mentally alert members of the community.'[27] Medical examination results certainly supported this view. During the course of the war, over three-quarters of all men aged between eighteen and nineteen who were examined by civilian medical boards were placed in Grade I, compared to only 40 per cent of those aged between thirty-six and forty.[28] Nevertheless, as available stocks of manpower became depleted, a second National Service Act of 1941 extended the upper age limit from forty-one to fifty-one years, bringing 2,750,000 more men under review.[29] Men in this older age group would be examined in same way as those who were younger but has the letter X added to their classification. Those in the highest grade would be graded, A(X)1, for example.[30] Yet this alteration in standard did not necessarily reflect a changing perception of fitness.[31] While older bodies could now be recruited, they were still considered as limited in what they could do. Winston Churchill explained in parliament that 'Men called up over the age of 41 will not be posted for the more active duties within the Forces. They will be used either for static or sedentary duties to liberate younger men.'[32] Although the needs of manpower had again rendered physical selection open to change rather than static, it did not mean that perceptions had been altered. This point is further supported by the fact that no men over forty-five years old and only a few over forty-one were ever enlisted.[33]

This idea of shifting standards while maintaining existing ideals can also be seen in the aesthetics of the body, particularly size and shape. Recruiting regulations before the war had stipulated that to join the army a man had to be a minimum height of 5 foot 2 inches, with minimum weight and chest sizes that varied according to height and age, but were no less than 8 stone and 33 inches.[34] While elite regiments such as the Guards and the Rifles continued to maintain strict height and weight requirements, no such minimums were set for the general militia, which represented the vast majority of recruits.[35] Nevertheless, it appears that examiners often paid much attention to the body's dimensions. Alexander Frederick volunteered for the army and was examined by a medical board in Dunfermline in 1939. He recalled that the doctor exclaimed, '"My God you're a big, fine looking laddie," and we were in.'[36] Roy Bolton was called up in and examined in London in 1942. He remembered that 'they looked very suspiciously at my chest. One of them said to me, "Did you ever have rickets as a child?" which I felt was slightly insulting. I was really pigeon-chested in those days.'[37] A Mass Observer, who was examined in London in 1940, also recorded the dialogue between himself and his examining doctors:

Doc I: 'Are you Grade I?'
Obs: 'I don't know.'
Doc I: 'Could you carry a pack and a rifle?'
Obs: 'I feel fit but I don't know how much they weigh.'
Doc II: 'You see you're very thin. Is your family thin?'
Obs: 'Some of them.'
Doc I: 'Ah I suppose you take after the thinner ones.'
(The two doctors went into conference. Apparently they couldn't reconcile thinness with fitness)
Doc II: 'In view of his eye perhaps he'd better go in Grade II.'
Doc I: 'We'll see what we can do for you.'
Doc II: 'There's always this problem with the athlete who doesn't look the part.'[38]

This medical encounter was not an objective assessment of the body but reflected wider social and cultural assumptions surrounding physical normality. This was based on the man's physical size, appearance and his lifestyle. The examiners focused on the fact that he was 'very thin'. Even though he stated that he 'felt fit', the doctors could not equate this with his small physique and so it confused them even more. In the end, the examiners were not simply classifying his body, but were using it to explore their own theories.

Examination

Bodies of the nation

Upon the introduction of the Military Training Act in May 1939, Minister of Health Walter Elliot made a presidential address to the Congress of the Royal Institute of Public Health and Hygiene in which he described the importance of the military medical exam:

> It will be an audit, so to speak, of the success or failure of the health services affecting the lives of citizens up to that age. The infant welfare services, the school medical inspection, the closing of the gap between school medical inspection and entry into health insurance, factory legislation, and all the housing and public health acts of our time will be brought into sharp focus – namely what is the health of John Smith at this crucial period in his early adult life, and what is the health of the 300,000 other John Smiths who came forward along with him?[39]

The data gathered from military medical exam results had long been used as a gauge for measuring national health. During the Boer recruitment campaign, between 40 and 60 per cent of men were rejected as unfit for military service. These statistics had caused increasing concern about the state of the population.[40] Anxieties about physical degeneration were further heightened during the First World War when 42 per cent of men examined by National Service medical boards had been placed in grades III and IV.[41] In light of these results, the government had implemented a range of policies during the interwar years that were designed to improve the national physique. As Elliot explained in his speech, these included provisions for better housing, child welfare services and industrial health reforms.[42] The examination of Britain's men for military service in 1939 was therefore not just another opportunity to monitor the condition of the population. It would be a measure of the success of the state's intervention in the nation's health.

When only 7.5 per cent of the first 50,000 men examined under the Military Training Act were rejected as unfit for service, politicians and other commentators were quick to take capital from these results.[43] According to Minister of Labour Ernest Brown, the high numbers of men passed as fit were the result of 'social legislation in the past thirty years, which had improved the health of the population'.[44] An article in *The Times* entitled 'An A1 nation' suggested that 'there is no reason to suppose that this first sample of young men is not representative of the country as a whole.'[45] Echoing this enthusiasm, Sir Ronald Davison, an expert on unemployment who worked for the Ministry of Labour during the war stated in a later article in the newspaper that the youth of the nation was 'healthier in every way than it was 20 years ago'. He continued:

> The explanation is doubtless complex. Poverty in its most abject form has been eliminated. Standards of cleanliness and clothing have immensely improved. Healthier habits and love of fresh air and exercise have become general among the youth of both sexes, even the poorest. It is much to their credit. Behind all this levelling-up lies one obvious and potent cause: the cumulative effect of the public social services built up since the turn of the century.[46]

Others, however, were less optimistic. Rather than viewing medical examination results as an indication of improved national health, some politicians and medical professionals considered them simply as evidence of a decline in recruiting standards. Critics drew attention to the fact that the pre-war regular army had a significantly higher rejection rate that the wartime militia. Recruiting statistics from 1936 had, for example, revealed that 22.6 per cent of men had been refused enlistment on medical grounds.[47] Highlighting this difference in the official figures, the *Lancet* reported in 1939 that 'The triumphs of social legislation can hardly explain the discrepancy between the acceptances for the militia today and for the army only three years ago, and suggestions have been made in parliament that the examination of the militia is inadequate and (provocatively) that "under any circumstances the men have got to be inside the armed forces".[48] Labour frontbenchers in particular drew attention to the fact that there were no formal height and weight requirements for the wartime militia and that while regular soldiers had been required to have a certain number of teeth, militiamen would not be 'rejected on dental grounds'.[49] Questions were also raised about the inclusion of men with defects of vision and of feet in Grade II, which many felt had lowered standards too far.[50] In the Commons, the Labour MP for Doncaster, John Morgan, complained that a man in his constituency had been passed as fit despite wearing a glass left eye. The representative for Durham, Joshua Ritson, similarly asked whether the Minister of Labour was aware that a young man in Sunderland had been passed as Grade II, even though he had only the sight of one eye and was under treatment by a specialist for the other. Ernest Brown simply reiterated that the classification system allowed such individuals to placed in Grade II as long as the remaining eye was sound. This prompted George Griffiths, Labour MP for Hemsworth, to ask, 'Is it not the fact that the Ayes have it up to now?'[51]

Examination

Sceptics also attributed the differences in rejection rates between regulars and militiamen to the types of individuals recruited in peace and war. While conscription imposed liability for military service on men of all classes, it was alleged that regular soldiers had tended to come from the poorer sections of society, which had fewer opportunities outside of an army career. In a tone of caution, the *Lancet* reported, 'Those who draw conclusions as to the improvement in the national physique by comparing figures for the militia with those for the army, and especially those for the army several years ago, are deceiving themselves. No one will deny that the health of our young men has turned out better than expected, but we must look elsewhere for proof that it is much better than it was.'[52] In a letter to *The Times* Dr H. Charnock suggested that 'They [the regular army] come on average from a poorer class than those taken from all classes for Militia conscription. One cannot but be struck by the great variation in physique, especially in height, between English public school boys and boys from the board schools of a similar age, amounting probably in the case of 15 year-old boys to three inches in height.'[53] Clearly, medical examination results did not alleviate all public health concerns but were seen as evidence of the continued division between the rich and poor.

Indeed, a closer investigation into the results gathered under the Military Training Act, conducted after the war, revealed that there were noticeable variations between men of different occupations and regional backgrounds. W.J. Martin from the MRC's Statistical Unit studied the examination records of over 90,000 men aged between twenty and twenty-one. He demonstrated that while there were no great differences in exam results between one part of the country and another, there were significant difference in the figures gathered from urban and rural areas. For the country as a whole, 84 per cent of men from rural districts were classified as Grade I, compared to 81.4 per cent of men from urban areas. Variations were greater still within specific geographical regions, especially industrial areas. In Glamorganshire and Monmouthshire in Wales, for example, 83.8 per cent of men from rural districts were placed in Grade I, compared to 76 per cent of those from urban areas. In Lancashire these figures were 84 and 77.4 per cent, respectively.[54] Details of the heights and weights of individual examinees also revealed further differences in physique between men from urban and rural backgrounds. In all regions of the UK, men from rural districts were heavier than their urban counterparts and in over three-quarters of all regions chest measurements were greater among rural dwellers. The greatest discrepancy in physical size was found to be in Scotland, where men from the rural Highlands were on average 1.1 inches taller, 9.2 pounds heavier and had an extra inch on their chest circumference than those from the industrial centres of Glasgow, Motherwell and Paisley.[55] Rural districts also had 10.6 per cent more men with perfect vision than those from large towns and cities. The biggest difference noted in this respect was again in Glamorganshire and Monmouthshire in Wales, where 72 per cent of men from rural districts were found to have perfect vision, compared to 60 per cent in urban areas.[56]

These statistics temper the optimism expressed by commentators in 1939 that low medical rejection rates were 'representative of the country as a whole'. Rather, in an era of widespread discussions surrounding the condition of the national stock, it seems that medical examination became a focal point around which wider social concerns could be expressed and political agendas could be furthered. While some government ministers and medical professionals privileged medical exam statistics as an objective reality in order to suggest that health inequalities had all but been eliminated since the First World War, others questioned the criteria by which men's bodies were being judged and continued to voice anxieties about the state of the urban population.

Detection and surveillance

Distrust of medical examination results in the early stages of the war did not centre only on the standards and classifications being applied to men's bodies. Some politicians and civilian doctors criticised the provisions of the exam itself, which they claimed did not allow for a proper assessment of the body. Questioning the four- to five-minute time frame that was allocated to each man, Labour MP for Leyton West, Reginald Sorensen, asked the Minister of Labour in July 1939 whether 'in those circumstances could there really have been a very thorough assessment of the militiamen?'[57] Subsequently, as the average age of men examined increased, the target was reduced from thirty men per two and a half hours to twenty-five in 1940, and decreased again to twenty-two in 1942.[58] More sophisticated techniques of bodily surveillance were also called for, particularly the inclusion of x-ray to identify men suffering from pulmonary tuberculosis, and psychiatric testing in order to detect those with nervous disorders. Together, these were the two greatest causes of manpower wastage among younger men in the army.[59] When, for instance, a man suffering from tuberculosis had been passed as fit on four separate occasions by civilian medical boards and army medical officers, the MP for the district commented that 'miniature radiography would have shown the heart to be drawn across to the right, and would have revealed coarse mottling of moderate extent behind the right clavicle'.[60] Highlighting the lack of adequate psychiatric screening, in April 1940 the *Lancet* reported that 'physically the recruit is being examined thoroughly enough...but much more could be done to weed out the mentally unfit'.[61] A month later the chief officer of the South East Lancashire Mental Welfare Association also even suggested that registered 'mental deficients' were being passed into the army and that psychiatric testing was the best solution to this problem.[62]

These debates were driven primarily by financial concerns and focused specifically on the subject of war pensions. The experience of the First World War had highlighted the financial burden to the state of compensating men for war-aggravated illnesses such as nervous disorders and tuberculosis. By 1929 over a million ex-servicemen had applied for pensions for 'shell shock related' disorders. By 1932 over 36 per cent of all British veterans who were receiving disability payments were claiming for psychiatric conditions.[63] In light of these statistics, the army psychotherapist T.A. Ross described psychiatric disorders as 'a serious problem for those responsible for the prosecution of the war, and an unending burden to the community'.[64] Another doctor claimed in the *Lancet* that nervous disorders had been 'a most desirable complaint from which to suffer because of the disability pensions that could be claimed'.[65] This opinion was echoed in the case for x-ray by Dr J.E. Bannen, who stated, 'I am surprised that the authorities have not insisted on such examinations, and it seems to me that they are carrying a very heavy responsibility for the future in the payment of pensions in cases which, if there had been an x-ray examination, such expense would have been avoided'.[66]

Ultimately, the demands for manpower overrode these concerns. In 1940 the government set up a Standing Committee on Tuberculosis to consider the inclusion of x-ray in the exam. It estimated that even if only one case were detected in every 1,000 it would be nearly ten times cheaper to use x-ray than the cost of treatment and pensions.[67] Nevertheless, a Medical Advisory Committee set up under the chairmanship of Lord Horder rejected the introduction of x-ray as impracticable, because it meant large numbers of men travelling longer distances to a smaller number of centres at which apparatus could be provided. It reported that although x-ray was 'an ideal to be aimed at in the medical examination of recruits for the Armed Forces', its introduction 'would seriously delay the examination of recruits'.[68] The same committee also considered, but rejected on four occasions, the introduction of psychiatric testing, recommending instead that 'an expert's opinion should be acquired only in cases of doubt.'[69] Explaining this decision in parliament, the Minister of Labour stated that 'it would be impracticable to include a doctor with such experience on every medical board.'[70] While acknowledging that examination was not flawless and was something to be improved in future, the state's priority in wartime was to maintain an adequate flow of recruits to fill the ranks.

In addition to these discussions about inadequate screening methods and technologies, some politicians and medical experts questioned the competency of the examiners themselves. In August 1939 the Medical Practitioners' Union journal *Medical World* alleged that 'the examination of recruits had been placed in the hands of men who had no experience of the work required'.[71] In October, Labour politician Thomas Groves also asked in parliament whether the Minister of Labour 'is aware of the increasing dissatisfaction among members of the medical profession with the personnel of the medical boards for the examination of recruits'.[72] These allegations were supported by several cases of errors in medical grading from around the UK. In October 1939 the Scottish Socialist politician and MP for Shettleston in Glasgow, John McGovern, brought to attention the case of Arthur Jamieson, who had been graded A1 despite suffering from fibrosis and arthritis. McGovern stated, 'the members of the board only looked at him in a casual manner and failed to subject him to a proper examination.'[73] In March 1940 George Hall, the Labour MP for Colne Valley in Yorkshire, raised the case of a man who had been in the army for three weeks before it was realised that he only had part of his right foot. Another man had been graded A1 despite having a missing right eye and receiving compensation for a hand injury sustained at work. Hall stated that 'a man in that condition was not really fit and it would not have taken even a doctor to see that.'[74] This issue was summed up by Samuel Silverman, the representative for Nelson and Colne, in 1942:

> People are saying that a medical board examining a recruit is interested only in getting him into the Forces by any reasonable means it can use, and therefore, if there is any doubt about a man's medical condition, the Army gets the benefit of it. It is said that the medical board come to the conclusion that any weakness will soon be brought out by the exigencies of military service, that the man in question will then be released and that they are prepared, therefore, to take a risk – and they take it.[75]

Examination

Under pressure to supply men for the forces, doctors were therefore accused of either not scrutinising bodies thoroughly enough or simply disregarding the provisions of the exam. Clearly, the mass screening of male bodies in wartime raised many questions about the broader fitness of the population, what actually constituted fitness for service, the adequacy of inspection procedures and the expertise of medical staff. It was not, however, only politicians and medical professionals who disputed the classifications applied to men's bodies during the war.

Contested bodies

The military medical exam undoubtedly highlighted the authority of the state to organise and monitor men's bodies in wartime and was a process that was designed to fit these bodies into a formal system of categorisation that left little room for self-identity or reflection. Certainly, for some men, it signalled the loss of control over their own bodies, as they were observed and analysed under this medical gaze.[76] Returning to the experience of the Mass Observer at a recruiting office in London:

> The doctors went over the same points again and again – for example, they all wanted to know why obs' right eye was weak and the left ok, why he had had a gland taken out, and why he was thin. In fact obs got the impression that they were treating him as a biological curiosity – he was very thin, and therefore must be unfit. Obs had to explain that his right eye had been injured by a tennis ball; that he was too young to remember anything about his gland; and that he had been active in a sporting sense. This apparently only seemed to mystify them even more.[77]

Although they asked him for his opinions of his own health and fitness, the examiners disregarded the answers that they were given, which only seemed to 'mystify them even more'. Despite feeling fit and being active, the man was viewed by the doctors with confusion and suspicion because of their notions of size and dismissed as a 'biological curiosity'.

Examination also constituted something of a performance, as men were observed while they completed certain tasks in order to provide the data for doctors to analyse. For Charles Lord, who volunteered for the territorials in September 1939, this compulsory visibility impeded his body's very functioning. He recalled his struggle to provide a urine sample in front of his examiner:

> I could never pass water to order and [laughs] I went and found out at this medical board that the senior guy was a doctor I knew and I couldn't do

this. I had to go and take my shoes and socks off and stand on a cold floor and okay it was there in the end. And of course he went home and told his wife and the first thing [laughs] it blew around Derby you see, silly [laughs] the first thing I was tackled with when I got back to Derby, 'Oh have you passed water alright all the way during the war?'[78]

Faced with the prospect of having to urinate 'to order', Charles's nerves seem to override his conscious control over his bodily functions. This, in turn, disrupted the examination process, prompting the doctor to make him stand bare-footed on a cold floor in order to collect the required sample and continue the examination.

The testimonies of men who entered into this sorting process reveal, however, that examination was not simply a moment of medical authority over the individual.[79] Many examinees tried to manipulate these medical encounters for their own purposes, particularly to avoid or secure enlistment. In this respect, the body can be seen as a site of contest and negotiation between the civilian and the representatives of the state. For those wishing to dodge military service, failing the exam was perhaps the best option. Ignoring one's call-up notice was a serious offence and carried the threat of imprisonment. Between 1939 and 1945 the Ministry of Labour authorised legal proceedings against 6,107 men who had not submitted to medical examination. Convictions were obtained in over three-quarters of these cases.[80] The government was also reluctant to accept refusal to enlist on the basis of conscientious objection. Of the 60,000 men who applied to local tribunals for such exemption, only around 3,500 were given unconditional exemption and over 12,000 had their claims dismissed outright.[81] Within this context, many men tried to be rejected for military duties on medical grounds. One strategy for doing so was to hire a man who had already been refused to take the exam for you. In June 1940 eight men in London were sent to prison after paying another man who was suffering from heart disease to impersonate them at medical boards.[82] In December of the same year two men were found guilty by a court of assizes in Stockport after one had paid the other, who had previously been placed in Grade IV, to attend the exam in his place.[83] Another tactic was to acquire false medical documentation. In December 1940 Arthur Carford, a member of the communist organisation, the Fourth International, was sentenced to eighteen months in prison. While working as an orderly at the Ministry of Labour medical board offices in Sheffield, he had stolen medical cards and planned to send them to the organisation in order to help members avoid conscription.[84] Civilian doctors could even be complicit in these attempts to thwart the medical

Examination

boards. During the war, several medical professionals were struck off the medical register for selling bogus medical certificates to men so that they might evade military service.[85] One doctor in London was paid between £200 and £250 each time.[86]

Other men put their own bodies at the heart of strategies of resistance by malingering. This meant either deliberately inventing symptoms of physical or mental disorders, or exaggerating symptoms that were already there.[87] A Mass Observer in Stepney in 1940 reported a conversation among a group of men in which one told the others that he had been graded C3. This prompted reactions such as, 'Here come on, out with it. How d'you do it? It's my turn next – got to get used to the idea. What d'you say?', and 'Bloomin' swindler – what'd you say to them? Told 'em you had a weak heart?'[88] A man who was examined in June 1940 also recorded in his diary:

> My own case attracted much attention. I was fit in every way, 4" chest expansion, etc., except for my unfortunate skin trouble. This, as I had expected, beat them all, though they all looked learned and gave it long names! I convinced them of its seriousness and they rated me 'GRADE III', which means I am not likely to be called up for a considerable time, if at all.[89]

In this instance, the examinee was able to negotiate the reaction of the doctors to his advantage, playing upon their desire to appear informed in order to avoid immediate enlistment.

Fooling examiners was not always so easy, however, despite the imaginative efforts made by reluctant recruits. In the *British Medical Journal* Captain Rankine Good of the Royal Army Medical Corps (RAMC) illustrated the case of one man who had turned up for his medical examination with his ear canal 'plugged with face cream in an endeavour to feign deafness'. The doctors quickly noticed this and asked if he was simply 'swinging the lead'. The man confessed that he 'had a horror of bloodshed and for that reason had by his malingering wished to avoid conscription'.[90] A Mass Observer in Crookham in 1940 reported a case of a man who had known his medical examination was at 7.30 a.m. and so stayed up all night drinking in nightclubs. By the evening after his exam he was 'almost passing out and kept saying tearfully, "they passed me A1 fit, can you believe it?"'[91] While these stories show that not all attempts to malinger were successful, they are, nevertheless, important. They highlight a perception of the medical exam as, in theory at least, something that *could* be traversed, and the body as both the site and the means of that negotiation.

In the same way, men who were keen to serve, but who were physically substandard, tried to manoeuvre around the exam's requirements. Corinna Peniston-Bird suggests that the medical exam did not only constitute a rite of passage into, or a moment of exclusion from, the armed forces. The classification of their bodies affected men's sense of their value to the nation, and their self-worth.[92] Certainly, a sense of wanting to belong in the higher medical categories is evident in the testimonies of wartime examinees. After his examination in London in 1940, one man exclaimed, 'They put me in Grade II! Pronounced me A1 and then put me in Grade II because of my vision! Impertinence I call it. However, I shan't go in the Front Line – that's one way of looking at it. It means clerical work.'[93] Despite being pleased about the overall outcome of his exam, this man's low medical grade was still experienced as a source of annoyance and embarrassment. In his diary, another man expressed the relief that he felt after being graded A1:

> I've been more than somewhat worried – far more than I'd admit, even to myself. It was in a way an unconscious worry, leading to a temporary inferiority complex which at times was almost unbearable. I felt out of everything I would be passed Grade 2, perhaps lower. I would be more or less an outcast. I wouldn't be able to look a woman in the eye etc...I feel that a great load has been lifted off my mind. I feel as if I've fallen in love.[94]

This account highlights the impact that wider hegemonic discourse could have on individual men, at a time when popular notions of masculinity were based on combat service.[95] The man in the story above stated that even if he had been recruited as Grade II, he felt that he would have been an 'outcast' and not 'able to look a woman in the eye'. It was not simply his inclusion in the armed forces but the role that he occupied once he got there that was crucial to his sense of self.

In order to avoid rejection many eager examinees tried to conceal their illnesses and disabilities. Again, one option was to pay a healthy man to take the exam instead. In 1941 the *Manchester Guardian* reported two such cases. One man had an artificial leg. The other was 'too blind to follow an occupation'. Both had paid another man of a similar age and build ten shillings to impersonate them at medical boards.[96] Others took their chances with the medical examiners and could be aided by the lack of adequate screening equipment. Dr A. Scott, a clinical tuberculosis officer in Wolverhampton, noted in a letter to the *British Medical Journal*, 'All over the country are men anxious to serve, among whom are a certain number of sufferers of pulmonary tuberculosis. With misguided patriotism these men are apt not to disclose their true history to the medical examiners, with the result that, lacking universal screening, many are missed clinically and passed into the services.'[97] It was also common practice for intending recruits who were rejected at one centre to try at other centres until they were accepted.[98]

Doctors could also be complicit in allowing men to enlist whatever their physical condition. As we have seen, it was alleged by some MPs that medical boards were simply passing unfit men in order to fill the ranks. Examinees' accounts also suggest that bodies were inspected in a casual or minimal way. Conscript Dennis March was medically examined in Leicestershire in October 1939. He recalled that 'my resulting medical examination confirmed I had a body, a head, two legs, two arms, my height was 5ft. 1in., my weight was 7 stones 5lbs, and declared me A1 "fit for anything"!'[99] A Mass Observation respondent who was examined in London in 1941 wrote in his diary, 'I can understand the stories of half-dead chaps being passed Grade I. The exam didn't seem so strict as it ought to have been.'[100] This potential for leniency on the part of medical boards could make it easier for men to negotiate the selection process if they wanted to enlist but were not physically suitable. Albert Parker was a nineteen-year-old county council clerk who described himself as 'all skinny' when he was called up in September 1939. He remembered the medical examiner asking, 'Oh, I don't know, do you really want to go?' When Albert responded, 'Of course I do. All my mates have passed and I want to go with them,' the doctor replied, 'Well, all right, yeah,' and he was passed as fit.[101] A certain amount of control could even be handed over to men in the classifications of their own bodies. Despite the instructions to medical boards stating that 'no discussion whatsoever, either in relation to disabilities or grading should take place within the hearing of any man under examination,' the experience of Eric Middleton shows that this rule was not always obeyed.[102] He recollected in an interview:

> The doctor who examined arms and legs turned out to be Dr. Rotherham. The family doctor we had when I was a child and who had said that I would never walk without irons. When he saw me he said 'Aren't you Alice Davidson's boy?' (my mother's maiden name). He had me doing various exercises, particularly going up and down on my toes with my arms extended to the sides, and called over the other doctors to show them what a miracle there had been. I was stark naked, of course, and there were some titters from my colleagues. He said he could grade me A1 (fit for overseas service) or B2 (for home duties) whichever I wished. I elected to be classed A1.[103]

Rather than being an objective assessment of his body, Eric's medical exam became a very public process involving many parties. Although the doctor made efforts to adhere to the formal structure of the exam by making Eric perform various exercises, ultimately Eric chose his own grade. This is a process Silbey refers to as 'adhering to the form while undercutting the intent'.[104] This was also the case when it came to the recording of external marks. Eric recalled that 'When a doctor came to identification marks he dictated "large mole on right buttock 2 inches from anus." I protested that it would be rather embarrassing to have to show that every time my identity was checked. He agreed to substitute "long scar on back of left leg"'.[105] A direct negotiation therefore took place, as both examiner and examinee played a part in shaping the medical 'facts'. As the result of Eric's protest, the doctor altered his analysis, but was again careful not to disregard the provisions of the exam. This moment of cooperation thus allowed medical science to be manipulated, as both parties were able to manoeuvre round the exam's requirements.

Conclusion

The military medical examination of the Second World War placed the body at the heart of the early stages of the transition from civilian to soldier. Whether volunteer or conscript, each man became subject to a formal sorting process in which his usefulness and personal future were decided by reference to his body. Examination determined whether each man was to be admitted to the army and what his potential as a soldier was at this early stage. This was reflected in the medical grading system, which not only sorted the generally fit from the unfit, but took account of various minor physical defects, abilities to see and hear and tolerance to climate. In grading this way, the authorities sought to make more effective use of all available bodies by suiting the right man to the right job. Through examination the body was also depersonalised. It was compared to ideals of usefulness and appearance linked to wider norms related to productivity and aesthetics, rather than judged on its individual merits.

Data from army examinations were also used for wider discussions of health. In an era dominated by anxieties over the condition of the British race and increasing public health reform, the statistics gathered by medical boards became a gauge for measuring national health. On the one hand, some government ministers used the militia's low rejection rates as evidence that the physical condition of the population had improved since the First World War. On the other hand, some doctors and MPs, particularly Labour ministers, who were still in opposition, looked sceptically upon these figures. Rather than seeing medical exam results as evidence of improvements in national health, they attributed them to the different social composition of the militia as compared to the peacetime regular army, whose higher rejection rates were felt to be a more accurate depiction of the condition of the working class.

The militia's high pass rates were also viewed as evidence that standards had been lowered too far in order to fulfil the demands of manpower, or that medical boards were simply allowing unfit men to pass. Indeed, at this early stage it was civilian doctors rather than military professionals who had the power to define the body's usefulness. In this context, the medical exam became a site of conflict and means of political point-scoring, as different parties held to different standards and ideals. There were clear differences of opinion among politicians and medical professionals about the standards adopted and the methods of bodily surveillance used. These debates centred mainly on the detection of tuberculosis and psychiatric disorders.

Examination was, therefore, a context in which the civilian body was observed, scrutinised and recorded by many 'experts'.[106] However, it also offered considerable opportunities for the men whose bodies were assessed. In this respect, the body itself became a site of both contestation and negotiation between the individual and the state. Some examinees who wished to enter the army but who did not meet the required physical standards consciously tried to hide their illnesses and conditions, while others who wished to evade service tried to cheat the selection process by malingering or getting others to impersonate them. Neither did doctors strictly adhere to the exam's requirements. While appearing to maintain some scientific impartiality, they could be lenient, even allowing men to choose their own physical grades. As such, examination was not an objective, detached scrutiny of the body but was something that could be manipulated by both sides. For the bodies that passed through this gateway, the process of reform was next.

Notes

1 R. Whitting, 'Civilian medical recruiting boards', in MacNalty (ed.) *The Civilian Health and Medical Services, Volume I*, p. 358.
2 Hansard HC Deb., 2 December 1941, vol. 376, col. 1030.

3 Ministry of Labour and National Service, *Manpower: The Story of Britain's Mobilisation for War* (London: HMSO, 1944), p. 11.
4 D. Silbey, 'Bodies and cultures collide: enlistment, the medical exam and the working class, 1914–1916', *Social History of Medicine* 17:1 (2004), 65, 75.
5 In October 1939 practitioners aged between forty and fifty were allowed to serve on medical boards, provided that at least two-thirds of the members of a board were over fifty. In areas where there was a shortage of available practitioners some men under forty were appointed. F.A.E. Crew, The Army Medical Services, Administration, Volume I (London, HMSO, 1953), pp. 348–9.
6 IWM SA, 20201, George Cozens, reel 1.
7 In May 1941 the number of examiners was reduced to three as it was felt that many of the duties of Examiners 2 and 4, such as taking height and weight measurements and testing urine, could be carried out by a medical orderly. Crew, The Army Medical Services, Administration, Volume I, p. 351.
8 TNA WO32/4726, Ministry of Labour and National Service, 'Instructions for the guidance of medical boards under the National Service (Armed Forces) Acts', 1940, pp. 2–3.
9 Crew, *The Army Medical Services: Administration, Volume I*, p. 330.
10 Ministry of National Service, *Report No 1 Upon the Physical Examination of Men of Military Age by National Service Boards from Nov 1st 1917–October 31st 1918* (Cmd. 504), p. 2.
11 Hansard HC Deb., 22 June 1939, vol. 348, col. 2432.
12 Crang, *The British Army and the People's War*, p. 5.
13 'Standards of fitness', *Lancet* (8 July 1939), 83.
14 Crew, *The Army Medical Services: Administration, Volume I*, p. 331.
15 Crew, *The Army Medical Services: Administration, Volume I*, p. 332.
16 Between June 1939 and October 1942 the percentage of men placed in grades III and IV almost tripled. Hansard HC Deb., 22 June 1939, vol. 348, cols 2427–34; Whitting, 'Civilian medical recruiting boards', p. 363.
17 Crew, *The Army Medical Services: Administration, Volume I*, pp. 345, 350–1.
18 Rabinach, *The Human Motor*, p. 277.
19 *Reports of the Industrial Fatigue Research Board, No. 16: Three Studies in Vocational Selection (General Series No.6)* (London: HMSO, 1922), p. 3.
20 B. Musico, 'A – The psycho-physiological capacities required by the hand compositor', in *Reports of the Industrial Fatigue Research Board, No. 16*, pp. 5–37; B. Musico, 'B – The measurement of physical strength with reference to vocational guidance', in *Reports of the Industrial Fatigue Research Board, No.16*, pp. 38–77; E. Farmer, 'C – Physical measurements in a sweet factory', *Reports of the Industrial Fatigue Research Board, No 16*, pp. 78–86.

21 Crang, *The British Army and the People's War*, p. 10; French, *Raising Churchill's Army*, p. 68.
22 R.S.F. Shilling, 'Industrial health research: the work of the Industrial Health Research Board, 1918-1944', *British Journal of Industrial Medicine* 1:3 (1944), 145-52.
23 C.S. Myers 'Introduction', in C.S. Myers (ed.), *Industrial Psychology* (London: HMSO, 1943), pp. 7-15; Crang, *The British Army and the People's War*, p. 10.
24 TNA WO32/4726, Instructions for the guidance of medical boards, p. 4.
25 TNA WO293/25, Army Council Instruction (ACI) 1428 of 1940, Instructions for the medical classification of soldiers by categories, 23 November 1940.
26 C. Peniston-Bird, 'Classifying the body in the Second World War: British men in and out of uniform', *Body and Society* 9:4 (2003), 33.
27 'Standing Committee on National Expenditure', *British Medical Journal* (20 September 1941), 410.
28 Whitting, 'Civilian medical recruiting boards', p. 361.
29 Hansard HC Deb., 2 December 1941, vol. 376, col. 1030.
30 TNA WO293/25, ACI 714 of 1942, Instructions for the medical classification of soldiers by categories, 4 April 1942.
31 Peniston-Bird, 'Classifying the body in the Second World War', 42.
32 Hansard HC Deb., 5 November 1940, vol. 365, col. 1928.
33 G. Forty, *British Army Handbook, 1939-1945* (London: Chancellor Press, 2000), p. 6.
34 TNA WO32/4643, Minimum height, weight and chest measurements, 4 July 1936, p. 13.
35 'Standards of fitness', *Lancet* (8 July 1939), 83.
36 IWM SA, 19804, Alexander Frederick, reel 1.
37 IWM SA, 23195, Roy Bolton, reel 1.
38 Mass Observation Archive (hereafter MOA) Topic Collection (hereafter TC) 29, Forces: Men in the Forces 1939-1956, 1/D, Report of JA of his medical examination, 3 August 1940, p. 3.
39 'Health services test: inspection of militiamen', *Scotsman* (24 May 1939), p. 16.
40 J.M. Winter, 'Military fitness and civilian health in Britain during the First World War', *Journal of Contemporary History* 15:2 (1980), 211.
41 Bourke, *Dismembering the Male*, p. 172; Winter, 'Military fitness and civilian health in Britain during the First World War', 212-15.
42 See above, Introduction; Jones, *Social Hygiene in Twentieth Century Britain*, p. 29; C. Webster, 'Healthy or hungry thirties', *History Workshop Journal* 13:1 (1982), 110-29; Constantine, *Social Conditions in Britain*, p. 35.
43 'Medical examination of militiamen', *Lancet* (1 July 1939), 48.
44 'Medical examination of militiamen', *Lancet* (1 July 1939), 48.

45 'An A1 nation', *The Times* (20 June 1939), p. 15.
46 Sir R. Davison, 'Youth's assize: the physique of the militia, a debt to an old war', *The Times* (1 August 1939), p. 13.
47 TNA WO32/4643, 97/Gen/9582, Report of Committee on Physical Standards, 1936, p. 4.
48 'Standards of fitness', *Lancet* (8 July 1939), 83.
49 Hansard HC Deb., 2 December 1941, vol. 376, col. 1030.
50 'Standards of fitness', *Lancet* (8 July 1939), 83.
51 Hansard HC Deb., 22 June 1939, vol. 348, cols 2430–2; 'Particular defects', *Lancet* (1 July 1939), 49.
52 'Standards of fitness', *Lancet* (8 July 1939), 83.
53 H.H. Charnock, 'Physique of the recruit', *The Times* (28 June 1939), p. 10.
54 Martin, 'The physique of young adult males', *Medical Research Council Memorandum 20* (London: HMSO, 1949), pp. 12, 22.
55 Martin, 'The physique of young adult males', p. 15.
56 Martin, 'The physique of young adult males', pp. 13, 63.
57 Hansard HC Deb., 22 June 1939, vol. 348, col. 2432.
58 Whitting, 'Civilian medical recruiting boards', p. 351.
59 Sir Arthur Salusbury MacNalty and W. Mellor (eds.), *Medical Services in War: The Principle Medical Lessons of the Second World War* (London: HMSO, 1968), p. 163.
60 John Aspin, 'A medical board deceived', *British Medical Journal* (5 October 1940), 470.
61 'Psychiatry and the services', *Lancet* (4 May 1940), 839.
62 Hansard HC Deb., 17 April 1940, vol. 359, col. 180.
63 E. Showalter, *Hystories: Hysterical Epidemics and Modern Culture* (London: Sage, 2003) pp. 73–4.
64 T.A. Ross, 'Psychological casualties in war', *British Medical Journal* (4 November 1939), 925.
65 'Neuroses in wartime', *Lancet* (31 October 1939), 1278.
66 'Correspondence', *British Medical Journal* (30 September 1939), 702.
67 TNA WO32/4726, Report on the radiological examination of recruits, 19 January 1940, p. 1.
68 The committee considered and rejected the introduction of x-ray first in 1940 and again in 1942. *Report of the Medical Advisory Committee on the Use of Mass Miniature Radiography in the Detection of Pulmonary Tuberculosis among Recruits for H. M. Forces* (1942, Cmd. 6353), p. 6.
69 TNA WO32/4726, Instructions for the guidance of medical boards, p. 4.
70 Hansard HC Deb., 27 February 1940, vol. 357, col. 1910.
71 'Examining recruits: medical journal Suggestion', *Scotsman* (14 August 1939), p. 15.
72 Hansard HC Deb., 19 October 1939, vol. 352, cols 1056–7.
73 Hansard HC Deb., 17 October 1939, vol. 352, col. 712.

Examination

74 Hansard HC Deb., 14 March 1940, vol. 358, cols 1447–8.
75 Hansard HC Deb., 29 April 1942, vol. 379, col. 1044.
76 Foucault, *Discipline and Punish*, p. 187.
77 MOA TC29, Forces: Men in the Forces 1939–1956, 1/D, Report by JA of his medical examination, 3 August 1940, p. 1.
78 IWM SA, 18257, Charles Lord, reel 1.
79 C. May, 'The clinical encounter and the problem of context', *Sociology* 41:1 (2007), 29–30.
80 Ministry of Labour and National Service, *Report for the Years 1939–1946* (1946–47, Cmd.7225), p. 22.
81 Ministry of Labour and National Service, *Report for the Years 1939–1946* (1946–47, Cmd.7225), p. 25.
82 'Evading army service: eight men sent to prison', *Scotsman* (29 June 1940), p. 6.
83 'Avoiding military service', *Manchester Guardian* (19 December 1940), p. 10.
84 'Conspiracy to avoid conscription', *Manchester Guardian* (19 December 1940), p. 7.
85 TNA WO32/4726, Medical Advisory Committee, Abuse of medical certificates, 3 July 1940, p. 2.
86 'False medical certificates: doctor struck off register', *The Times* (27 November 1942), p. 2; 'Men aided to avoid military service: doctor struck off register', *The Times* (30 November 1942), p. 2.
87 Major J.A. Hadfield, 'War neuroses', *British Medical Journal* (28 February 1942), 284.
88 MOA TC29, Forces: Men in the Forces 1939–1956, 1/B, Observations of NM in Stepney, 29 July 1940.
89 MOA D 5006, Diary for June 1940, p. 4.
90 Capt. R. Good, 'Malingering', *British Medical Journal* (26 September 1942), 360.
91 MOA TC29, Forces: Men in the Forces 1939–1956, 1/F, Comments about soldiers collected by DH in Crookham, p. 1.
92 Peniston-Bird, 'Classifying the body in and out of uniform', 35.
93 MOA TC29, Forces: Men in the Forces 1939–1956, 1/D, Report by JA of his medical examination, 3 August 1940, p. 1.
94 MOA D 5039, Diary for February 1941, pp. 7–8.
95 Rose, *Which People's War*, p. 179.
96 'One legged man: how he joined up', *Manchester Guardian* (19 February 1941), p. 7.
97 'X-ray examination of militiamen', *British Medical Journal* (23 December 1939), 1250.
98 See, for example, 'Medical examination of militiamen', *Lancet* (5 July 1939), 104.

99 BBC WW2 People's War Archive, A1109161, Dennis March, 14 July 2003, www.bbc.co.uk/history/ww2peopleswar/stories/61/a1109161.shtml (accessed November 2013).
100 MOA D 5175, Diary for May 1940, p. 4.
101 IWM SA, 14788, Albert Parker, reel 2.
102 TNA WO32/4726, Instructions for the guidance of medical boards, p. 7.
103 BBC WW2 People's War Archive, A5610485, Eric Middleton, 8 September 2005, www.bbc.co.uk/history/ww2peopleswar/stories/85/a5610485.shtml (accessed November 2013).
104 Silbey, 'Bodies and cultures collide', 8.
105 BBC WW2 People's War Archive, A5610485, Middleton.
106 Foucault, *Discipline and Punish*, p. 304.

2

Training

Once enlisted into the army, every new recruit underwent a period of basic training. At the start of the war this generally lasted for between three and four months and was carried out in depots, before men were posted to regiments.[1] After the introduction of the General Service Scheme in July 1942, recruits spent their first six weeks in newly created Primary Training Centres, where they underwent basic infantry training, aptitude and intelligence tests. They were then posted to Corps Training Centres to receive instruction specific to their arm of the service. This lasted for between sixteen weeks for infantrymen and a maximum of thirty weeks for signallers.[2] The purpose of all of this was to prepare recruits for active service according to two key principles. Upon being confronted with their new intake, army staff had to establish control over each man's body in order to submit him to the authority of the regime. Instructors also proceeded to transform the body into an effective military machine by rendering it fit, ordered and ultimately self-regulating. The army manual *Basic and Battle Physical Training* stated that 'the battlefield is the supreme test of training. When units are divided into small and scattered groups each man must be held responsible for his own fitness. Every individual should have the will to be fit, and every man must be taught what exercises he can do under the particular conditions existing at the time. He should realize that only if he keeps himself fit can he be an efficient soldier.'[3] This chapter thus explores the various techniques by which the civilian body was adapted for military utilisation by focusing on the two main principles in this process: control and transformation.

Civilians into soldiers

Controlling the body

The very first step in the army's training regime was to establish control over the recruit's body in order to create an effective basis from which the process of reform could take place. This began the moment that he arrived at barracks, where he would be stripped of his civilian identity through the issue of an army number and uniform and by being given the regulation haircut. Scots Guards recruit Peter Grant arrived for basic training in Chelsea in 1940. He recalled in his memoir, 'we went to the barber, who completed each cut in about a minute. The electric machine rendered the hair very short and only a longer frill or "dosan" at the front was left to make something of.'[4] Eric Murray underwent his basic training at Brancepeth Castle in Durham in 1943. He remembered 'young fellers with a mass of curly hair and they did this short back and sides. They went straight over the top and you looked more like a skin head.'[5] Some men clearly felt a sense of pride in these outward manifestations of military identity. Henry Novy trained at an RAMC depot in Leeds. He wrote in his diary in November 1940 that 'the pride taken in the uniform is that of being part of a well-organised and efficient machine.'[6] Bill Partridge, who was conscripted into the Royal Ordinance Signal Corps in 1939, also recalled in an interview, 'when I came home I was a little bit different to my contemporaries. I was in uniform.'[7] Others, however, were more upset about the immersion of their individual bodies into the generalised mass. On his second day at Queen's Barracks in Perth in 1942 one infantry recruit recorded in his diary that 'by this time we were dressed in our denims (work suit) and I for one felt even more depressed. We sent all our clothes home and now the complete break from civilian life was accentuated.'[8]

After this initial introduction to military life, everything about the recruit's appearance continued to be subject to strict control and regulation. He was to be kept clean and smart and observe meticulous attention to detail in the wearing of his uniform. This included caps to be worn over the right side of the head with the front peak one inch above the centre of the right eyebrow, creases in the trousers, folds over the anklets, and boots were to be highly polished.[9] Men were also expected to shave daily, which sometimes had no physical or aesthetic value. Upon arrival for basic training with York Rifle Brigade in 1942, eighteen-year-old William Dilworth explained to his sergeant major that he had never shaved before, as he had never even had any fluff on his chin. William later recalled the sergeant major's response:

> The sergeant major looked at me and he said, 'don't tell me what you haven't got.' He says, 'I'm telling you. You need a shave. Corporal,' he called one of the corporals and he said, 'take this man to the ablutions and see that he shaves immediately.' So the corporal marched me off across the square to the toilets and he said, 'Well, shave,' and I said, 'I've never shaved.' So he says, 'Get your razor out,' so I got the razor, which the army gave me and opened it up and I said, 'Well, I haven't got a razor blade,' so he says, 'Put it all back together again and go through the motions, soap your face and then make out you're shaving.' So, without a razor blade in I went through the motions of shaving and everything, washed my face and was marched back to the parade ground, marched up to the sergeant major and the sergeant major looked at my face and said, 'That's bloody better, man. Now in future you'll shave every morning.'[10]

The very fact that Dilworth, like all men in the army, was issued with a razor suggests that shaving was an expected behaviour, regardless of whether or not it was necessary. It was not designed simply to standardise men's appearances but to enable the authorities to instil discipline, order and routine through a particular habituated bodily practice.

It was not just the exterior body that became subject to the army's strict control. In order to be made efficient, all of the body's functioning first had to be regulated. As such, the body was conceived as a machine to be fine-tuned:

> If the engine is to continue to function efficiently all the necessary supplies must be maintained. The petrol must be ample, and a free air assures, together with accurate timing of ignition and sufficient cleaning of the exhaust. So, in the body, break down of any one part, though it can be compensated to a degree not obtainable in a man-made machine, has its effect on the efficient functioning of the whole body. Ample supplies of food and drink, free access of oxygen from the lungs by blood transport to the tissues, accurate timing of nervous control of each individual muscle fibre, and efficient cleansing of the waste products through the venous blood. All these and many more complicated details are necessary for efficient functioning of men in training.[11]

Thus, the body was to be fed, rested and cleansed of its contaminating waste products. The manual *Physical and Recreational Training* advised that, among the 'foundations of physical fitness' were 'good nutrition, which is achieved by careful attention to the soldier's diet' and 'careful regulation of drinking and smoking'.[12] Men in training were to be provided with 'a substantial meal at the beginning of the day, followed by two lighter, easily digested meals at a four or five hour interval, ending the day's work with a fourth good substantial meal'. Daily provisions included specified amounts of meat, fish, vegetables, pulses and cereals. These were 'designed to cover all requirements, including body-building proteins, energy-producing carbohydrates and fats, and protective vitamins and salts'.[13] Recruits were also to secure eight hours of sleep at night and adequate periods of rest throughout the day, including every meal being followed by half an hour of physical relaxation in order to help the 'digestive processes get underway'. This, especially in the evenings, allowed for 'the opening of the bowels'.[14]

While designed to ensure the efficient functioning of the body in order to produce a soldierly performance, this emphasis on diet and digestion also reflected wider discussions about health that had been at the heart of a range of social policy interventions during the early twentieth century.[15] The Inter-Departmental Committee on Physical Deterioration had identified food as fundamental to health and proposed reforms such as free school meals and 'social education' for housewives who lacked sound knowledge of nutrition.[16] There had also been increasing professional interest in the science of nutrition during the interwar years, with the emergence of several state-sponsored and company-owned nutritional laboratories. According to James Vernon, researchers 'sought to discover the precise thermodynamic laws governing the body and the exact chemical properties of food'.[17] With the outbreak of war in 1939, feeding the civilian workforce also became a primary government concern in order to stimulate good health and, consequently, industrial production. Under the Factory (Canteens) Orders of 1940 and 1943, the number of workplace canteens rose from under 6,000 in 1941 to over 11,500 by the end of the war. Factory canteen advisers were appointed to instruct employers on matters such as cooking, menu planning, nutrition, staffing and layout.[18]

The army also drew on this wider knowledge of nutrition in order to feed the new recruits. The War Office employed several civilian experts in dietetics, including a special advisory committee on nutrition, as well as specially trained messing officers and advisory catering officers, who were attached to the various commands.[19] Attention was paid not only to nutritional intake, but also to providing a pleasing diversity of food to the men. The *Manual of Military Cooking and Dietary* advised that the sergeant-cook 'will afford every facility for varying the diet of the several messes, so that each mess may have a complete change daily throughout the week'.[20] To this end, twenty-four army cookery schools were set up where civilian catering experts trained army cooks in matters of both nutrition and variety. The *Manchester Guardian* reported in December 1941, 'We are doing our best to give army cooks, who in past days used to be regarded as men not fit for any other job, a definite and respected status. The trained Army cook will provide variety in the daily menus, not the perpetual stew.'[21] The importance of 'careful supervision of meals' was even more evident in the army's Physical Development Centres. These 'special reconditioning camps' were established during the war in order to improve the physical categories of men who had been placed in grades A2, B1 and B2, thus 'raising their efficiency and rendering them fit for employment in a more active and strenuous capacity'.[22] Recruits received 1 kilogram of food, or 4,000–5,000 calories per day, made up of 250 grams of water and roughage, 500 grams of carbohydrate and fat, 120 grams of protein and various vitamins and minerals.[23] Instrumental in the creation of the centres was Adjutant-General Ronald Adam, who, as Harrison notes, 'was deeply committed to health and education for the masses, and saw both as vital to the successful prosecution of a "people's war"'.[24] The results gained were certainly encouraging. Eighty-one per cent of recruits sent to the first centre opened in Kingston-upon-Thames were raised in medical category at the end of the course.[25]

Indeed, there was feeling among recruits from various regiments that this management of their bodies was beneficial, especially in light of rationing and food shortages in the civilian population. Walter Chalmers, a twenty-one-year-old volunteer in the 1st Battalion, Liverpool Scottish at Boughbridge, believed that 'the food was very much better than one would be getting in civilian life at that time.'[26] Ordinance Corps conscript Bill Partridge also suggested that men in his barracks 'felt better. They were getting adequate meals. They were getting sufficient food and there was a sense of wellbeing.'[27] This is not to say that all men were happy about the meals they were served. Roy Bolton remembered that some of his fellow recruits 'really moaned and groaned' and 'complained bitterly about what I found perfectly edible and was really quite enjoying'.[28] Yet even men who found the diet somewhat plain, or were used to eating more, confessed a sense of trust in the military regime. Recruited at eighteen years old, Joe Stevens trained with the Royal Artillery in Woolwich. He later recalled, 'bearing in mind that most of us were young boys, we were permanently hungry and we thought that they didn't give us enough food. But everybody's health and physique improved under the training.'[29] In a report of a conversation among a group of recruits about the meals served at his depot in Leeds, Henry Novy also recorded one man as saying, 'We reckon we get good food here. You know, not fancy stuff, but good wholesome grub. Yesterday I saw some people lift their noses but they'll get used to it. I've seen many a bloke come here with a pale face and go away as right as rain.' Another man agreed, stating, 'They know what they're doing here. At home you overeat.'[30]

Amidst fears over the spread of venereal disease among both the military and general population, the army also sought to establish control over the recruit's body by regulating his sexual desires and behaviours.[31] The pamphlet *The Soldier's Welfare: Notes for Officers* advised that 'War confronts the civilian turned soldier with many new and complexing problems, but none perhaps so urgent or so difficult as those concerned with his sex behaviours.'[32] This meant that contact with women was often strictly monitored. At William Dilworth's barracks in Retford, for example, there was a dance every Saturday night that local girls were invited to. Promptly at ten o'clock, all the women were put into army lorries and taken back to town. William remembered that 'they were all checked and everything to make sure nobody was staying behind.'[33] Many soldiers also suspected that bromide was added to their tea in order to, in one man's words, 'keep your sexual fantasies down.'[34] It is likely, however, that such stories reflect a perceived element of bodily control rather than a reality. The belief that the authorities used drugs to tame men's libidos was an enduring myth within the military.[35] In actual fact, it was probably the change of lifestyle and the strenuous regime that recruits now had to follow that caused them to experience a reduced sex drive. This was certainly the opinion of Ron Gray, who trained at Bulford Barracks on Salisbury Plain in 1941. He stated in interview that 'everybody believed this universal myth in the armed forces that the cook put something in the tea. It's because sexual appetite just diminishes because there's no sort of sexual provocation around you.'[36] Leonard England recorded in his diary in April 1941 that 'sex is one of the things that ceases to matter so very much. In a discussion the other night between all types in the hut it was agreed that this could not fully be explained by bromide in the tea. To some extent, it was due to the utter change in living conditions.'[37] A Mass Observation respondent named Morris, who trained at an RAMC depot in Essex in 1941, likewise noted in his diary, 'Those who find that the comparatively active and varied pattern of army life while in training leaves little energy for working their sex glands – in this connection may be mentioned the current legend that the cooks are instructed to add bromide, in something similar to the tea, as so many men notice their own lessened sexual activity, both in sexual coition and in masturbation.'[38]

The question of masturbation as a suitable alternative to sexual intercourse was in fact raised among the medical community. In 1944 a doctor wrote into the *British Medical Journal*, asking whether it would be 'true and right' to tell a man in the forces that masturbation was a lesser evil than fornication, and what the ill results of masturbation were. The journal responded that:

> He can be assured that occasional masturbation can be resorted to without danger to his health or to his future sexual life. If he decides to adopt this measure he should be told to adopt it deliberately, knowing that it is an emergency measure undertaken deliberately and after careful consideration. No sense of guilt should be attached to the act, and it should be resorted to only when it seems essential to obtain relief for a sexual tension which has become unbearable. If masturbation is used in this way and not merely for pleasure, it will be unlikely to become a preferable substitute for normal intercourse in the future. He must be told also, that the ill effects that are popularly supposed to follow masturbation are usually due to excessive masturbation.[39]

This suggests that even masturbation was something to be carefully regulated, carried out only as an 'emergency measure' and not simply for pleasure. The recommendation that it should be adopted 'deliberately' also implies that it was a technique directed towards self-control as it enabled soldiers to manage their own sexual needs and desires.

It was not only heterosexual behaviours that the military authorities tried to police. As in civil society, homosexual acts were illegal in the armed forces. Classified under the offence of 'indecency', an ordinary soldier found guilty of this crime could face up to two years' imprisonment, while an officer risked being cashiered (dismissed).[40] Yet, while the peacetime army could be more selective about the sexual preferences of the men it recruited, the needs of manpower during war meant that many homosexuals were now absorbed into the ranks.[41] In order to prevent illicit sexual encounters, army leaders adopted several techniques. Joseph Inskip trained with the Coldstream Guards in Caterham in 1940. He explained in an interview that men were not allowed to sit on each other's beds because 'homosexuality was absolutely out'.[42] Henry Butterworth, a trainee at a depot in Berwick-upon-Tweed, recalled that any men suspected of 'hanky-panky' were given a cold shower.[43] The sergeant major in charge of Bill Partridge's unit at Chilwell also warned one particular recruit that 'I've got my eye on you. Keep yourself to yourself. I don't want to say any more, but if I ever see you squatting down to pee I'll kick your arse.'[44] As such, his efforts to monitor the man's sexual behaviour focused on his body itself as a material marker of sexuality. He believed that his homosexuality would manifest itself in a particular bodily trait, 'squatting down to pee', a typically feminine performance.

This linking of homosexuality with corporeal attributes had dominated both popular and medical conceptions of sexuality since the late nineteenth century and had led to a range of scientific studies that tried to connect physical characteristics to homosexual desire.[45] During the war such ideas quickly permeated the military arena. In the United States, for instance, the armed forces screened for homosexuality by relying on physical bodily markers, like a 'feminine distribution of pubic hair and fat deposits' and 'effeminate gestures and mannerisms'.[46] In the UK, psychiatrist Charles Anderson likewise conducted a series of studies on both 'active' and 'passive' homosexual soldiers at Wharncliffe Neurosis Centre. He concluded that while there was 'no measurable deviation from the physical normal', around three-quarters of the passive homosexuals 'presented a slightly feminine appearance'.[47] Recruits also located sexuality with discernible feminine qualities. Kenneth Bond, a private in the 2nd Battalion Gloucestershire Regiment remembered one man who 'you didn't want to be too close to. He was much too effeminate.'[48] Douglas Arnold, a recruit in the 7th Commandos in Ipswich in 1940, recalled that 'people were called pansies in those days, but to be a pansy you had to dress with a silk scarf, a flower in your buttonhole, things like that, wearing funny clothes. We never thought of pansies sleeping together.'[49] Douglas seems to have had a clear idea of what a gay man should look like: someone who used feminine props, like silk scarves and flowers. It actually surprised him to think that men slept with each other. For him, appearance, rather than choice of sexual partner, defined sexuality.

All of these processes of regulating the recruit's body – its appearance, nutritional intake and sexual activities – were accompanied by close surveillance of his body in order to monitor his conformity to standard. The body was routinely maintained through regular inspections and parades, which allowed any defects to be identified and remedied. Private Joseph Clarke recalled, 'you just used to strip off and the MO [Medical Officer] used to ask you if there were any problems or owt like that.' He also remembered 'infection inspections', which were 'mostly round your private parts and so on'.[50] Again, this scrutinising of the body began from the moment that men arrived at barracks. William Dilworth, who had a tooth 'drilled and filled', explained that 'you go through a thorough medical examination, your ears and every other part of the body and,

and then…everything was okay cos you have all these things done there and then so you go before the doctor and he checks everything and then he says, "Right, dentist," so you walk straight over to the dentist's chair and have it done there and then.'[51] Examinations and parades were not, however, simply about imposing control from 'outside'. Rather, prescribed bodily behaviours became routine, ingrained practices, as men came to expect to be inspected. Weekly kit inspections, for instance, made sure that recruits prepared their uniforms in strict accordance with army standards. Bill Partridge commented that 'somebody was going to be stupid if you hadn't made up your uniform and bed space or whatever so get on with it and okay we worked out the routine probably a lot quicker.'[52] This monitoring of men's bodies was reinforced through systems of reward and punishment. For recruits on guard duty the 'stick' or 'fetch-and-carry man' was often selected on the basis of personal cleanliness. William Dilworth was granted this privilege as a result of scrubbing under his fingernails:

> We're all standing there at attention and the regimental sergeant major comes along and says, 'Hands behind your backs.' So we all put our hands behind our backs and he marches along in front of us and he looks at every one of us, inspects all our uniforms and beard and everything, you know, to see if we'd shaved, haircut and what have you. Then he goes round the back and…when he got to me he lifted up my hands. He'd obviously done it with all the others…when he'd finished doing that he went to the front. He said, 'Right, step forward Rifleman Dilworth.' So I stepped forward one pace. He said, 'You will be the man that looks after the rest of the men,' which means I have to run backwards and forwards to the canteen, get drinks and food and everything for them while they're on duty. So I didn't have to stand and do any guard at all. I got away with it because I had clean fingernails.[53]

Men who were not up to the required standard of smartness, on the other hand, risked being confined to barracks.[54] Joseph Clarke, a private in 9th Battalion Durham Light Infantry recalled that 'there was always someone getting CB [confined to barracks], always.'[55] Dilworth also explained that men who did not look after their kit and maintain good standards of personal appearance were 'put on a "you can't go out tonight" or something like that. Your leave was cancelled for the night.'[56] The intended effect of this was that men would be discouraged from reoffending. Frederick Cottier, a regimental policeman at Fenham Barracks in Newcastle, explained, 'we always found that once a man had been sentenced to CB or detention, he behaved himself.'[57] Punishments were also essentially corrective and designed to move the body towards self-improvement. Recruits who failed uniform inspections due to unpolished boots or a generally untidy appearance, for example, were not only confined to barracks but were made to perform additional nightly parades in which their standards of dress would be re-examined.[58]

The body was also controlled in time and space. Where the recruit could go, when he could go and how he could go were all subject to strict regulation. From the moment that he awoke until he went to sleep he experienced a routine existence regulated by a timetable. For some veterans this became ingrained in memory. William Corbould trained with the Coldstream Guards in 1940. He remembered that 'The working day started at 0700, followed by drill, breakfast, halls of study, lunch, afternoon parade and PT [physical training]. The working day then finished between 4 and 5 p.m., followed by dinner, then evening preparations, such as cleaning.'[59] Even the most elementary bodily functions were regulated in this way. Bill Partridge explained that 'ablutions were to be completed one hour after the morning stand down and that was for everybody. There was no question of, oh well I'll go to the latrine after the morning break or anything like that.'[60]

Men were also considerably restricted in their personal movements. Confinement to barracks was not just used as a form of punishment but was often the norm, particularly during the early days of training. Russell King attended an infantry training centre in 1940 and recalled that 'nobody got leave for the first sixteen weeks.'[61] Henry Novy reported from his depot in Leeds in December 1940 that many applications for day passes or weekend leaves were refused without explanation. His own application for compassionate leave in order to get married had even been rejected.[62] When men were allowed out of barracks, there were also still limitations on where they could go. One soldier noted in his diary in August 1940 that the men at his depot were 'confined to within a radius of about 5 miles…To get outside this area one can do so only by means of a day pass at the weekend of which there are four between thirty men.'[63] Within the enclosed military camp personal movement was further controlled through the 'partitioning' of space into functional sites, such as the canteen, barrack room and training ground.[64] Describing Reedsdale Camp in Nottingham, Lance Sergeant Ian Sinclair remarked, 'You had to go to the drill hall for your dinner. You had to go to the drill hall for your evening meal and you had to march back. You couldn't walk back by yourself when you were ready.'[65] The way in which space was allocated and utilised therefore allowed the authorities to continually monitor and

Training

observe each individual body as men were no longer free to move, eat and sleep as and when they wished.

Placed within the wider military organisation, the body also assumed a specific, though not fixed, position through the allocation of rank. This was inscribed on the body through dress. Rank badges and stripes were worn on the shoulder, and each particular regiment had its own idiosyncrasies that differentiated the officers from the men. James Allen Ford, an officer in the 2nd Battalion Royal Scots, recalled that 'We had to pay the regimental tailor to put a ribbon round about and dangling down the back, to distinguish us from the ordinary troops.'[66] The classification of bodies was similarly a spatial process. J.A. Bergin also noticed at Blanford Camp in 1940 that the NAAFI (the Navy, Army and Air Force Institute-run canteen or recreational club) was divided into two parts 'for men and Corporals (Bombardiers), each being forbidden into the other's half'.[67] Lance Sergeant Ian Sinclair also described the segregations of bodies at Reedesdale Camp in Nottingham:

> There was different food from the men had, you didn't have to go into the mess, you had your own mess tent and the food was served by orderlies, which was really something. Comparatively you were a thing apart as much superior, for want of a better word, as the officers were to the sergeants because the sergeants' mess was sacrosanct. The only people who weren't sergeants that got into the sergeants' mess were the orderlies, or when you had a sergeants' mess night the officers came into the sergeants' mess.[68]

Sinclair's testimony shows that the division of bodies went further than just officers and men: a clearly identifiable multi-tiered system was in place. As a sergeant, he was 'a thing apart' from the rank and file who were lower down the military hierarchy, but also separate from the officers, who were superior to him. As such, each group was designated its own particular space.

Transforming the body

Once harnessed, the recruit's body also became a site of transformation, to be made fit, ordered and productive. This began with basic physical training, a daily routine designed to create a base level of fitness and equip the body with the fundamental skills and abilities that were necessary for active service. In order to achieve this, instructors used combinations of exercises, all of which, according to *Basic and Battle Physical Training*, were 'designed to have a beneficial effect on some part of the body or to contribute towards the development of mental qualities'.[69] The five main principles that basic physical training was designed to develop were: mobility; strength; endurance; agility, dexterity and speed; and carriage. In order to achieve these, men took part in balance exercises, marching, running, skipping, heaving, climbing, vaulting and games.[70] Within every exercise, the body was broken down into individual motions and gestures. Each was assigned a specific speed, aptitude and direction in order to achieve the optimum results. *Basic and Battle Physical Training* stressed that 'it is important that the correct form and range of each movement is insisted upon by the instructor, otherwise the exercises will lose their corrective value and slack,' and that every exercise 'has a characteristic speed which will yield the maximum effect'.[71] For example, men were taught to relax all muscles not required in such movements as walking, running, crawling, climbing, lifting and pulling, in order to conserve a maximum amount of energy.[72] The army's strict focus on bodily form and movement can be seen clearly in the technique for road walking:

(i) Walking heel and toe (two parallel lines) toes to the front.
(ii) Free, relaxed action of the rear leg as it comes forward.
(iii) Arm action, arms bent to 90°, hands relaxed, passing close to body.
(iv) Upright carriage of body, full chest, head erect, looking straight ahead.
(v) Smooth progressive action, body does not rise and fall with each pace.[73]

Having acquired the fundamental bodily skills, recruits then progressed to battle training. This was designed to hone the body for warfare by teaching the application of these skills in more military activities. Exercises included 'forced marching, running, surmounting obstacles of the type likely to be encountered in the field or in street fighting, landing by parachute, jumping from tanks or other vehicles on the move, climbing and scaling, lifting and carrying, close combat, gun, mortar and truck man-handling, swimming, and crossing water obstacles by means of improvised aids'.[74] William Dilworth described an assault course:

> You'd be taken out by lorries to some rough part of the country and they'd say right there's paper trails and that you follow that, you know, and course it would be all over the place, the rough and tumble and everything like that and then course there was the actual field where they had all these tough things to climb over and swing from and all that sort of thing, over water and all that.[75]

The main principle behind all of this was progression – the gradual rising of each individual body to the desired level of proficiency. Lessons were 'always arranged according to the capabilities of the individuals for whom they are intended, gradually increasing in difficulty or severity from week to week and month to month, so as to ensure steady and systematic progress throughout the whole course of training'.[76] The distance covered in a route march was also gradually increased over time so that men could steadily build up their fitness. William Dilworth recalled that 'basic training began with a slow one-hour route march and culminated in a two-hour 10-mile march, performed at a speed of 180 paces per minute'.[77] Through the application of rigorous and systematic effort, men's bodies became standardised, as each individual was progressively made fitter, stronger and enhanced in stamina and endurance.

Physical training was not, however, designed simply to reform the individual body, but to render that body an important cog in the wider military machine. *Basic and Battle Physical Training* advised that every soldier should be able 'to reach his maximum potential skill in any physical activity required of him in his particular army duties. He will thus become competent to play his full part in the team work of his sub-unit in battle'.[78] In order to foster unit cohesion, the authorities adopted various bodily strategies, such as collective physical movement, which induced a sense of solidarity and a commitment to shared goals.[79] It was during a route march, when he fell in line with the other men that conscript Henry Novy suddenly experienced the feeling of becoming a soldier:

> This marching was queer – at the beginning I felt for the first time, almost in spite of myself, that pride in numbers, marching numbers, squad after squad in step. I saw it in many men's eyes, looking proudly to the passers by. They were happy to be carrying full kit and marching, squad after squad, over 400 men. When we had our kits on, two of my mates remarked: 'Here we go boys, real soldiers now'. The little coalminer said: 'You feel a proper soldier now, don't you?'[80]

Novy appears to have felt almost powerless to stop the changes that were happening to him as, physically and mentally, he was immersed into the collective body of men.

This was also the primary function of drill, a highly ritualised bodily ceremony in which men were made to perform precise movements and gestures to an external rhythm imposed through the call of commands. In his memoir, Peter Grant described 'a chorus of bawling sergeant majors flourishing their pace sticks to ensure that every man's step was of regulation dimensions'.[81] The intended outcome of this was that men would learn to direct their bodies to respond automatically to orders, thereby instilling the predictability of behaviour necessary for success in battle. Field Marshal Earl Wavell described drill as 'the outward, the mechanical side of discipline, learning by practice to do something so automatically that it becomes natural even in moments of stress'.[82] Ron Gray underwent basic training at Bulford Barracks on Salisbury Plain in 1941. He clearly recognised the army's objective when recounting his experience of drill. He remembered 'Miles and miles of endless marching. You stamp up and down and you march and you halt and you march and you halt and you march, as though it's designed really to crush your brain power [laughs], to turn you into an automaton. That seemed to be the point of army training, as though we weren't really human beings.'[83] The main effect of drill was, however, the coordination of mass mind and movement, as recruits became attuned to each other's roles and were orientated towards a common goal. Scots Guard Officer W.A. Elliot explained in his memoir:

> By drilling soldiers into a sense of uniformity, the members of each unit come physically to act and perceive themselves on parade as acting as one man. The suggestive patter of the drill sergeants was incessant. The first ten minutes of each drill period were expended in 'warming up' or 'chasing' in double quick time – left turn, right turn, about turn, halt! – and woe betides the man who ended up facing in the wrong direction. Some of the new recruits simply could not take all this 'chasing'...But gradually a growing sense of uniformity began to emerge and to outweigh the fear of retribution, on which less emphasis now needed to be placed.[84]

Less formal bodily cultures were also directed towards collective discipline. During physical training lessons, classes were divided into groups in order to inculcate 'the team spirit'.[85] The army also placed great emphasis on organised games and sports to develop unit cohesion and *esprit de corps*. Sport had become officially integrated into military life and training during the First World War. Drawing on the nineteenth-century public school tradition, military officials valued competitive games and sports as useful for improving fitness, relieving boredom, building morale and fostering officer–man relations.[86] Sporting activities continued to be central to the training syllabus during the interwar years and throughout the Second World War.[87] Recruits took part in interplatoon boxing, swimming, soccer, hockey, cricket, rope climbing and cross-country running. Again, these were valued both for their physical effects and for promoting 'comradeship' and 'wholehearted

Training

Figure 2 Infantry recruits undergoing rifle drill at Chichester Barracks, 1939

cooperation and an unselfish attitude for the good of the side'.[88] These qualities were made clear in the army's definition of a 'sportsman', who was described in *Basic and Battle Physical Training* as 'the man who takes all decisions without question or argument', who 'is unselfish and always ready to teach and help others' and, 'when a spectator, cheers good play on both sides, but never interferes with referees or players'.[89] Regular officers like the regimental sergeant major (RSM) in charge of Bill Partridge's regiment placed great faith in sporting proficiency as a measure of a man's character. One of the first things that Bill vividly remembered about basic training was the RSM asking, 'Right, rugger or soccer?' When Bill replied, 'I play rugger, sir' the RSM said, 'Good', and he was immediately selected for officer training.[90]

All of these processes of transforming men's bodies were again reinforced with various accompanying strategies in order to achieve the desired results. Recruits were offered incentives to encourage them to perform well, such as a free weekend pass for the man who came first in the weekly cross-country run or was selected to play rugby for the regimental team.⁹¹ Alternatively, those who failed to meet the required standards were usually punished. William Corbould made mistakes while on parade:

> If you were a naughty boy, as I was on two occasions, and got punished and confined to barracks, you then had to do drill on a Wednesday afternoon, Saturday afternoon and Sunday afternoon, and I mean drill at the rate of knots on the Old College square. You reported and away you went. On one occasion I had failed to swing my arm correctly in supernumerary rank, coming back from Old College to New College and so I was given four extra drills for being idle on parade.⁹²

Once more this punishment was corrective, moving William's body towards the desired result through an intensified regime of training.

Men who did not meet the demands of their training instructors were also publically humiliated. W.A. Elliot noted that during drill, 'when some individual made a wrong movement he would be snarled at and told he was not in the bloody so and sos.'⁹³ Masculinity could play an important role in this process, for a body that was unreformed was also constructed as unmanly. Physical training sergeant Ian Sinclair suggested, 'It was very hard and made some of them wish they'd never been born. To have to go on a three-hour route march killed them, falling out by the wayside, and you have to try again with the discipline, try and make them. It began to sort out the men from the boys at a very early stage.'⁹⁴ Such classifications were then internalised by recruits, whose own abilities to cope with the rigours of training became crucial to their sense of self.⁹⁵ On the day that he passed-out in the Coldstream Guards Joseph Inskip remarked to a senior officer, 'I'm a man now and I'm a trained soldier.'⁹⁶ As such, training was a rite of passage that signalled Joseph's transformation both from a civilian into a soldier and from a boy into a man. After his route march Novy also reported, 'When we came in, tired, all tried to say they loved it and felt no effects. A lad with bad feet dropped out. His mate remarked: "I'd rather be dead than drop out of a route march, I would honest". To my shame I must say I felt the same, a pride of being a soldier, well disciplined, in step, doing hard work.'⁹⁷ The labels that were assigned to bodies appear to have inculcated self-discipline in individual recruits. Rather than being compelled to march, the men in Novy's squad were driven by a sense of pride and a fear of humiliation. This meant that they laboured over their own bodies in order to maintain a particular self-identity.⁹⁸

The wartime army also, however, adopted a strategy of kindness and cooperation in order to induce men's bodies to transform. *Basic and Battle*

Physical Training advised that 'Instruction in physical training must be positive, never negative. Discouragement is worse than useless, since it causes a negative anxiety on the part of the man.'⁹⁹ Bill Partridge recalled this sort of approach in his own training. He explained that his physical training instructors 'were very sensible men and they took it to the stage that nobody said, "Touch your toes," and if, as I could never touch my toes ever, I was always two or three inches short, they would never say "Do it or else."'¹⁰⁰ This approach was deemed particularly suitable for conscript soldiers, who were considered by officials as 'different human material' from peacetime regulars. As men who had been employed as professionals and skilled workers, they were accustomed to using 'initiative, skill, intelligence and individual effort'. Consequently, they were more amenable to cooperation and explanation rather than more traditional forms of discipline.¹⁰¹ In order to inspire self-discipline, officers and physical training instructors were encouraged to interact with their recruits and to share in the hardships of training. *Basic and Battle Physical Training* advised, 'To create the desire for physical efficiency in his men the instructor must himself be the living embodiment of fitness.'¹⁰² In a lecture on 'man-management' in 1942 Lieutenant-Colonel R.A. Mansell of the RAMC also commented that 'It is rather surprising how many sports grounds one can visit and see the men playing hard, without ever an officer looking on. You do realize what a difference the personal interest of an officer does make to all of the men's activities, don't you – even his games? And how he does appreciate that an officer should play games with him?'¹⁰³ Through all of these additional strategies men were encouraged to engage in their own physical transformations, rather than the army simply imposing its designs upon them. In this way, officers and instructors could instil the self-regulation, the 'will to be fit' that was necessary for victory in battle.¹⁰⁴

The body responds

Soldiers' testimonies show that men responded in various ways to the army's efforts to shape and control their bodies. Importantly, they tell us that the army's ambitions for the body were not always realised. Some recruits, it seems, simply could not conform, as they struggled to cope with the rigours of military life. This was often the case during parade, when men found it hard to perform the precise movements and gestures demanded by their drill sergeants. Russell King recalled, 'Well, basically it was just turning, how to turn right, how to turn left and how to about turn. You know, those were the very basics of course. Oddly enough lots of people couldn't do that. There was quite a few lads found it very difficult, just sort of putting their feet right, you know, and that sort of thing, in order to turn.'[105] Joseph Inskip also remembered a fellow recruit in the Coldstream Guards who 'couldn't change step. That is a rule in the Guards. Change step and the whole lot changes step, and he just couldn't do it. He just could not do it.'[106] Such accounts convey the body's non-conformity as unintentional. Poor coordination was innate so the inability to march was not really the men's fault. Russell noted that lads 'found it very difficult' just to perform the moves. Joseph asserted that his fellow recruit 'just could not do it.' These stories suggest that the body's limitations were beyond the immediate control of both the army instructors and the men themselves.

Other recruits, however, intentionally resisted the demands of their officers and instructors. When he progressed to officer training with the Ordinance Survey Corps, Bill Partridge was advised by his mentor, 'if you get a group of thirty men, some are going to be keen, some are going to do as they're told, a few you will have to watch and kick.'[107] Physical training instructor Ian Sinclair also explained, 'Being told "you must," "you will," rather than, "will you?" That was the thing that was going to really get discipline going as such, and some people didn't take kindly to it.'[108] Resistance to the military regime was again considered to be a particular problem with conscript troops, many of whom begrudged their changed roles and responsibilities. Mass Observer A. Calder Marshall noted in a report on 'morale and training in the army' that 'There exists naturally a large unconscious resentment at being drafted out of a civilian job, with which they are familiar and at which they are earning good money, into military service with its low pay and attendant discomforts. Rather than a source of delight, army life is at first an unpleasant duty.'[109]

Efforts to resist the demands of training were not, however, confined to conscript recruits. While training with the York and Lancashire Regiment, infantry volunteer R.A. Graydon discovered one way of avoiding a particular disliked physical activity. He joined the battalion band as a clarinet player after discovering that this meant that he did not have to complete a full route march. He explained in his memoir that 'this lucky group of individuals would play the men off into the distance with a few rousing marches and return to camp. Three hours or so later, they would go out to pick them up again, hoping to be able to bring them to life again for the last half mile.'[110] Volunteer Percy Bowpitt had been so eager to join

Training

up that he lied about his age to get into the army in 1942. Enlisted into the Northamptonshire Regiment, Percy and his friends quickly came to realise how physically demanding basic training was, especially the dreaded weekly cross-country run:

> Once again we had to endure the weekly cross-country run beloved of the Army PT Instructors. Our route took us out of town, through farms and fields and back through the town. This had the advantage that when the edge of town was reached it was possible to hop on a bus…Provided the bus stopped some way from the barracks all was well but often the conductor would deliberately pass the stop we needed and then stop nearer to the barrack where would be standing Regimental Police waiting to catch anyone too slow off the mark.[111]

In response to the confinement of their bodies, some men simply left their camps altogether, either by deserting or going absent without leave (AWOL). According to military law, desertion demonstrated a desire to remain away permanently, while absence without leave was a temporary measure usually taken by men who wanted a break from their military responsibilities.[112] Soldiers who were discovered going absent were usually punished upon their return to camp, either with confinement to barracks, extra fatigues or detention in military police cells.[113] In order to avoid such penalties, men who went AWOL often used various tricks to conceal their absence. During his training with Durham Light Infantry at Brancepeth Castle Robert Ellison would leave camp to visit his girlfriend and get a friend to call his name at the morning register.[114] Percy Bowpitt also described dodging Military Police by 'climbing station fences, running along the track, waiting until their backs were turned and generally using the field craft that the Army had so usefully taught us. A major problem was returning to barracks without being caught.'[115] Deserters were more concerned with simply getting away and gave less thought to the consequences. In December 1940 Henry Novy reported that 'Four men have already deserted because leave had been refused. Many say "Oh it's daft, it doesn't get you anywhere, they always catch you in the end", but asked what they would do in similar circumstances they say they would leave immediately if refused.'[116] In response to this behaviour, however, military authority was simply intensified. Two weeks later, having been told that they would not get leave for Christmas, Novy reported that some of the men 'were just buggering off, apparently when they thought fit'. The sergeant major's reaction was that 'they were stupid blighters, and wouldn't dream of doing it if they had really got into the military discipline state of mind.' So he abolished all weekend leave under any pretext except compassionate leave.[117]

So, rather than trying to remove their bodies from the military environment, other men again put their bodies at the heart of strategies of resistance by malingering. In a diary entry for January 1941, Leonard England described 'a very strong resentment to P.T., which is held out in the open at 8.30 in vest and shorts. Over 50%, I should say attribute their coughs and colds and ailments to it and all sorts of excuses are used to get out of it.'[118] Henry Novy also noted that 'going sick is the right of every soldier, and as it involved hours of waiting it is one of the favourite ways of escaping official big parades.'[119] In 1942 Lieutenant-Colonel William Brockbank of the RAMC described how some soldiers tried to feign dyspepsia (upset stomach). He noted that 'There was a type case that complained of vomiting after every meal and yet contrived to look uncommonly fit on it. This vomiting was done without eye-witnesses and appeared to cease directly when the man was forced to vomit in public. These patients probably exaggerated their symptoms. Most of them had been to hospital once or twice, their story being more spectacular on each occasion.'[120] Captain Rankine Good also encountered several cases during the early years of the war, including a recruit who 'developed the symptom of a paralysed hand' but admitted that he was malingering after three weeks in hospital. Another soldier pleaded loss of memory, having given himself up for going AWOL.[121]

Such determination to resist by malingering, and the responses to it, were not new in the armed forces during the Second World War, but had a longer history in both the military and civilian spheres. Roger Cooter suggests that 'malingering and its detection was in and of modernity.' It was part of the 'routinized, disciplining demands of the modern industrial world and its warfare'.[122] The introduction of workmen's compensation schemes since the late nineteenth century had offered financial incentives to malinger by providing compensation for illness or injury sustained at work.[123] In this context there had been increasing concern about the socio-economic implications of malingering and consequently a surge of interest in it.[124] Books such as John Collie's *Malingering and Feigned Illness* and Arthur Bassett Jones, Llewellyn J. Llewellyn and William Mardon Beaumont's *Malingering: or the Simulation of Disease* provided advice on the detection of malingering by medical examination.[125] Joanna Bourke has shown that exposing malingering also became an important business for British military officials during the First World War. Between 1914 and 1918 servicemen tried to evade duties by feigning headaches, sleeplessness or dizziness, or simulating diseases like appendicitis and lumbago. Medical staff had thus adopted a range of techniques to detect offenders, such as careful examination of physical symptoms and looking for particular bodily signs like a look of 'open-eyed-candour', 'cunning' or a 'tendency to overreact'.[126]

With the outbreak of hostilities in 1939, the military doctor again became the detective. This role was usually played out at the regimental sick parade, which was described by one captain as 'a daily battle of wits between doctor and patient'.[127] Experiences like that of Harold Pollins suggest that this medical encounter was one based on suspicion and mistrust. Harold developed suspected septicaemia while at an infantry training centre in 1944. He recalled, 'I lined up with the others on the sick parade to find ourselves being inspected by the RSM. He inquired disdainfully into the reasons for our reporting sick and clearly did not believe the answers he got – the usual ones about flat feet, stomach ache and the rest. His tone suggested that he thought we were all malingerers.'[128] There was, however, no consensus on how to tackle the problem. Fearing the malingerers' potential to disrupt good discipline and morale, some medical officers favoured dismissing the men. Rankine Good considered the malingerer 'as real a threat as the potential infectivity of a typhoid carrier' and recommended the 'removal of the weak link in an otherwise strong chain'.[129] Another approach was to punish men who reported sick. At a barracks in Winchester recruits reporting sick were excused from normal training activities, but were often given extra fatigues such as cleaning or helping in the cookhouse. A Mass Observation report noted, 'at the present time, the attitude of many NCOs [non-commissioned officers] to men reporting sick is that they are malingering…If men are sick they should be treated as sick. If they are suffering from blisters or something which incapacitates them from going on regular activities, they should be learning something useful instead of picking paper out of dustbins that ought never to have been put there. The attitude of suspecting malingerers produces malingerers.'[130] Other medical staff focused on better methods of detection through more intensive bodily surveillance. One medical officer advised that all cases recommended for discharge should be transferred to a psychological depot 'where they could be kept under continuous observation before finally being boarded out of the Army'.[131] Major H.A. Sandiford of the RAMC also suggested that 'in some of these cases it is a most difficult task to decide when a man is malingering, and they usually require more than one examination. It is often necessary to have a man specially observed for some time.'[132] Not all medical staff were, however, happy about the army's expectation that they should act as detectives. In February 1942 Major J.A. Haddersfield of the RAMC warned against the dangers of wrongfully accusing innocent men. In an article in the *British Medical Journal* he expressed concern that too many medical officers were not discharging recruits with neurotic or psychopathic symptoms because they were 'swayed by prejudices' that these men 'should get away with it'. Describing this as 'a grave injustice', he stated that 'it is just as criminal to send a combatant officer an unreliable man of poor calibre who is likely to break down as it is to send him a defective gun or an untested plane'.[133] Yet even those who were sympathetic to the men felt under considerable pressure to police malingering. In November 1944 another army doctor wrote to the journal, describing the ethical tension that he felt between trying to maintain manpower and preserving individual health:

> His [the soldier's] position on sick parade is prejudiced from the start by the fact that one must put one's loyalty to the Service first. One thinks not of a patient to be healed but of [a] man who may be trying to get away with something (not a malingerer, for they are rare, but an exaggerator), and he has to prove he is ill before you start to think of him. Then in the background of our minds is a wild procession of forms to be filled in about him. Gradually, in spite of oneself, one ceases to be a doctor in the best sense, a healer, but becomes a State servant.[134]

Because reporting sick aroused so much suspicion, another option for the recruit was actually to inflict harm upon his body. During a period working in the medical inspection room in Lydd Barracks in Kent, trainee stretcher-bearer Frank Offiler took care of a man who had shot himself in the foot.[135] Rankine Good also reported the case of a recruit who was so fed up with the lack of social life in the army that, 'when shaving, he had made a great display of his razor.'[136] The ultimate act of reclaiming one's own body in this respect was suicide. Regimental policeman Frederick Cottier described how a man named Anderson tried to kill himself at Fenham Barracks in October 1939:

> Corporal Thornton got a call to say there was a soldier lying on his bed with his rifle pointing to the door with a round up the spout, threatening anyone who came in. Lance Corporal Thornton took the rifle from him and placed him under close arrest. That night I was on cells duty. It was my job to make sure that all the equipment was removed from his cell with the exception of his two blankets, as a safety precaution against attempted suicide. When I went into the cell I found Anderson hanging by the neck. He'd fastened his kit-bag rope from the bars and stepped off the bed. He

was unconscious. I lowered him down and went for assistance. The medical orderly performed artificial respiration on him. He recovered and was sent to the hospital.[137]

Infantry recruit Sam Beard recalled the successful suicides of two 'vulnerable points' (low medical grade) men at Norton Barracks in Worcestershire in 1939. They were discovered by a sergeant, who ran out of their hut and exclaimed that 'there's two vulnerable points blokes in there...They each...got between the sandbag and the hut, knelt down facing one another, put their rifles under each other's chin and shot each other.'[138] Such were the measures that recruits would go to in order to escape army life.

Other men did, however, find ways of coping with the changes in lifestyle imposed by military service. While adhering to official codes of behaviour during the working day, during off-duty periods recruits were able to engage in bodily pleasures and excesses. For example, although drunkenness was a crime and alcohol was forbidden in camp among the rank and file, soldiers' testimonies reveal that much free time was spent vising local pubs. Edward Kirby recognised the comfort that alcohol could bring to men who were apart from their loved ones. Recalling his experiences of basic training with the Royal Engineers in Andover, he explained that 'for some of the men who'd been separated from their wives and families, they were rather fed up, you know, and they lost their boredom and their sorrows, very often, like the ones I had, in drink. I don't blame them. I mean, I felt that I could have drunk gallons of whisky after I'd met them.'[139] It seems that these men used alcohol to deal with the loneliness and monotony of barrack life.

Despite the army's efforts to monitor sexual behaviour, recruits also found opportunities to satisfy their needs and desires. Russell King commented that 'there was plenty going on, of course, for young lads, of course, there were all the local girls and there was local dances.'[140] Another soldier wrote in his diary that 'for the first fortnight we were confined to camp and there were no dances: there was no sex life at all. After that, girls were available at the dances.'[141] While these accounts do not state explicitly that the men had sex with the girls, it is implied. Others chose to pursue relationships with servicewomen or civilian women who worked in their camps. Reporting from his depot in Essex, Morris described a group of men who 'keep their urges fastened on such females as may be within range – the girls in the canteen and those living in the neighbourhood'.[142] Leonard England suggested that recruits found servicewomen preferable. He noted that 'the ATS [Auxiliary Territorial

Service] are far more in demand as "girl-friends" than civilian women in the depot.'[143] Indeed, separation from loved ones could enhance men's sexual ambitions. One new recruit wrote in his diary in May 1941 that 'since being in the Army my feelings for my girl-friend have changed, especially my desire for her sexually. I have seen her three times since I have been away, and she has told me that I am "much more affectionate" than I was before.'[144] In order to cope with their frustrations, men who were confined to barracks regaled each other with details of their sex lives.[145] Leonard England noted in his diary that one form of 'sexual release' was to describe 'moments of passion' and read love letters to one another.[146] Reports from Mass Observers also highlighted the prevalence of 'sexual epithets' used in the army, in comparison to civilian life. This was due to 'lack of sexual satisfaction, no doubt accentuated by boredom and leisure time to be spent with little money and few if any civilian friends'.[147]

In the safety of their barrack rooms, men also had sex with each other. In his memoir R.H. Lloyd Jones described two recruits at Blanford Camp whose 'behaviour all day was perfectly proper. It was therefore a little surprising sometimes, when awakening in the morning to see the two of them sharing the same bed.'[148] Morris also commented on the openness of homosexuality at his depot in Essex:

> Apart from the posturings on the barrack room beds (sometimes almost, though never quite amounting the performance of the act itself) it is a quite well-recognised fact that such activities do occur, and that those who participate will freely admit to them. This may be due to the fact that recruits include all types, not omitting those already well versed in these arts. Here one must bear in mind the age (20–30), varied walks of life, from costermongers to university students, of the people concerned, and the outside opportunities for the "working off" of sex.[149]

On the one hand, it seems that military service forced these men to reorganise their sexual patterns in order to cope with their frustrations. Within the enclosed all-male environment, and with little privacy or free time, they turned to each other in order to satisfy their sexual desires. As Morris put it, there were few 'outside opportunities for the working off sex'. Certainly, similar relationships were observed among men in other armed services and all-male organisations, suggesting a link between homosexual behaviour and institutional culture.[150] On the other hand, with the influx of homosexuals into the armed forces during wartime, these men could simply have brought their outside sexuality into the military environment. As Morris noted, wartime recruitment brought in 'all types', including those 'already well versed in these arts'. So, it is quite possible that the men were not reacting to their military situation but were using their bodies in ways that were already familiar to them.

Moreover, these sorts of behaviours were not just about physical gratification. Drinking became something of a bonding ritual among new recruits and firm friendships were made over a pint of beer. Dick Fiddament, who trained as a young soldier in the Royal Norfolk Regiment, recalled that 'it was nothing to see five or six guys...men, pals, sitting around, all playing at having a game of cards, sharing a drink...I was closer to them than my own biological brother.'[151] Some military superiors also drank with their men in order to encourage bonding and *esprit de corps*. Charlie Workman, an officer in the Fife and Forfar Yeomanry at Chippenham Camp in Newmarket, regularly took his young recruits on a 'troop night out' to the local pub, where he bought the first round.[152] The ability to drink large amounts of alcohol and the ardent pursuit of women were also key features of macho culture within the ranks that young recruits tried hard to live up to.[153] This is clear in the experience of grammar school boy Roy Bolton, who, on his first night at Richmond Barracks, visited the NAAFI with some of his new friends:

> It was taken for granted that we'd all drink beer so one of us went out to get beer. To my horror a great foaming pint of beer was put in front of me and I'd never drunk a pint of beer before, no more that half a pint [laughs], a fact which I concealed and I set myself manually to drink this. It seemed to take an immensely long time to get through it and I was alarmed to find that my mates were getting through theirs more quickly. I tell you, I did get to the bottom of it and eventually, more or less within the six weeks we were there, I learned to beyond...I think the first time we had about two pints each, which was flooding my stomach. I think I got as far as four pints eventually, where I was slightly worse for wear.[154]

A desire to fit in with one's peers likewise accounted for the stories of sexual prowess that were so common in barrack rooms. One recruit wrote in his diary that 'It is difficult not to twitch from stories that are told in garbled versions after the event.'[155] Henry Novy reported that 'smutty stories go around quite a lot, often pure dirt without much point.'[156] Dick Fiddament also recalled, 'there was some near-the-knuckle kind of jokes, but truthfully, and I say this hand on heart, there was very little.'[157] A sense of competition was therefore induced as men engaged in performances in order to suggest that they were sexually promiscuous. More innocent recruits like twenty-year-old Ron Gray felt under considerable pressure to live up to these expectations. He recalled in an interview:

> Sex was just a dirty topic. I mean, if you talk sex, you talk dirty. It's the only way soldiers seem to be able to talk about it. I used to hear these hair-raising stories of sexual activities and I didn't...I was still a virgin. I didn't know, I didn't know how to measure up. It dawned on me how bloody ignorant I was about the whole subject. I mean, I had urges like the next guy, but apart from that.[158]

In order to conceal his inexperience, Ron ordered a sexual encyclopaedia that 'embraced everything about sex and aberrations and strange activities that were a revelation for me'. This allowed him to join in future conversations about sex.

Military superiors could also be complicit in bending the official rules. Three days before Christmas the corporal in charge of Novy's depot told the men, 'if we can't get out, let's get bloody drunk.'[159] In addition to alcohol being forbidden in barracks, was also an offence for men to return from off-duty periods intoxicated. Soldiers who ignored this rule and were caught usually faced two weeks' confinement to barracks. However, as the testimony of military policeman Frederick Cottier shows, official procedure was not always carried out. He manned the guard room at Fenham Barracks in Newcastle and was responsible for inspecting the men as they came in. He recalled that while some individuals were detained and sent before the corporal, there were 'six or seven particular characters' who 'were perpetual drunks but never came up on any other charge'. The guards put these men into the cells to sober up but released them in the morning. Frederick explained that 'we just used to release them because on the whole, let's face it, they were good soldiers, they had a night out, and we used to lock them up to prevent them from causing further disturbance, more or less in their own interest.'[160] Rather than treating all cases of drunkenness as the same, the guards were therefore selective in the men they chose to report. If a man was not considered a threat to military discipline and was 'a good soldier' he was allowed to sidestep the usual rules. Indeed, Frederick recalled how, aware that they would only be held in the cells overnight, some of the men even 'fetched their blankets with them' before leaving the barracks. Safe in the knowledge that they would not be reported, they were able to get drunk without fear of punishment.

Similarly, officers and men engaged in sexual liaisons with each other. Rupert Lyons, an infantry recruit in Pankhurst Barracks on the Isle of

Wight, described one encounter between a corporal named Smith and a soldier, Bobby Neal:

> One night after he had been out drinking, he came into our barracks after lights out. He put the light on and went around looking at people. He pulled back people's blankets to examine them. He pulled off my blankets, and had a look. I grabbed my blankets back and he went on to the next bed, then on to the other side of the barrack room. Then he came to the last bed where Bobby Neal slept. Bobby woke up, smiled then jumped out of bed. He was a great passive homosexual was Bobby Neal, just waiting for someone to come and take him up.[161]

While Smith's position of authority may have allowed him to enter the men's space and pursue a sexual encounter, it is clear that Bobby willingly complied. He did not wait to be 'examined' like the other men, but eagerly volunteered to go. Morris likewise observed sexual relationships between officers and men at his depot in Essex:

> There are a certain number who are definitely treated as females by the others. They are referred to by feminine pronouns, 'chi-ified' by the less inclined and use feminine first names (Sheila, Nora, Elsie). Some of these men go so far as to 'make up' in the evening with eyebrow pencil, rouge and lipstick, and a certain neighbouring public house is supposed to be the favourite haunt in fixing any rendez-vous. I am speaking now only of the private soldiers with whom I am most in contact. These men-women often refer avidly to their officer gentlemen friends and a certain few N.C.O.s are popularly supposed to utilise their services, but whether the other ranks practice similarly among themselves is more than I can say. Possibly not, as their opportunities for a more normal sexual other are far greater.[162]

Again, it is hard to determine exactly what drove this behaviour. As Morris suggested, it may have been a response to the all-male environment, which caused some soldiers to take on feminine identities and have sex with other men.

Yet it is also possible that the men were simply reflecting their prior attitudes and orientations. In his study of queer urban lives in early twentieth-century Britain, Matt Houlbrook has highlighted the openness of queer culture, which included characters like the 'queans' and 'Dilly Boys' of London's West End, who adopted feminine names, dressed as women and wore make-up. These were mainly working-class men who often slept with middle-class 'respectable homosexuals', not just for economic benefit, but as a way of asserting their deviance or of feeling empowered.[163] Indeed, it is telling that it was the private soldiers who made themselves up as women and chose to sleep with their officers. This suggests that the behaviour constituted something more than either the men's own existing dispositions or the desire to find a sexual outlet when opportunities were limited. Marjorie Garber suggests that cross-dressing has a long history within the military and attests to what she calls 'carnivalized power relations': the subversion of the power structure through the profane.[164] By making themselves up as women and having sex with their superiors, these recruits may have developed a creative or 'safe' way of mocking the military hierarchy and asserting their own power, if only temporarily.[165] The officers also could have had various reasons for participating in these relationships. They too could have been homosexuals in civilian life or they may have sought to reaffirm their position of authority by retaining the traditionally dominant male role. For whatever reasons, it seems that normal rules were temporarily suspended so that both sides could fulfil their own objectives.[166]

It is equally important to note that recruits could comply with the army's demands for their bodies in order to achieve productive ends of their own.[167] As we have seen, some soldiers enjoyed eating more and experienced a sense of pride in working hard and marching in step. Others confessed a sense of enjoyment and wellbeing through the physical training regime. One man noted in his diary in 1940 that 'I'm fitter than I was on civvy street.'[168] Another commented, 'I used to pride myself on my fitness as a result of training for cycle racing before I entered army life but this daily physical training has developed my stamina tremendously.'[169] Roy Bolton, who described himself as 'all skinny' and 'pigeon-chested' when he was recruited in 1942, also described his own bodily transformation through training:

> Practically everything we did seemed to be of a physical nature, to which I just wasn't accustomed to. I just had no occasion to do it. Very much the clerical type in the office and so it was that combined with copious supplies of food and the fact that I was always ready to eat the food, always hungry. So, lots of exercise, lots of food and I just got bigger and better, which did me a lot of good.[170]

These men seem to have decided that the hard work was a price worth paying for a body that looked and felt better. While training may have benefitted the army, it is clear that some recruits felt that it also benefitted them.

Conclusion

Every man recruited into the ranks of the British Army between 1939 and 1945 underwent a four-month regime of rigorous indoctrination that was designed to strip away his civilian identity and prepare him for all the elements of active service. This process of training the civilian to operate as a soldier directly and explicitly targeted his body.

New recruits faced an immediate onslaught of regulatory controls upon their bodies. Upon arrival at their barracks they were clothed, had their heads shaved, were medically inspected and were given any necessary immediate treatments. They then continued to be subject to detailed surveillance of dress, diet, hygiene, sexual behaviours and personal movement. Men were told how to wash and shave, what to eat, when to rest and when to work. Control even extended to the most elementary functioning of the body through the regulation of the guts. In order that these routines became more and more habituated, recruits were constantly inspected and rewarded or punished depending on their conformity to standard. All of these techniques allowed military instructors to achieve authority over the bodies and the minds of the men who were sent to their units.

The body was also a site for reform, to be made ready for all the duties required in active operations. Through a strenuous regime of basic physical training, battle training and drill, instructors attempted to transform each body into a more predictable and effective fighting machine. As well as increased stamina, strength and endurance, recruits learned to obey orders instantly, to understand the chain of command and to put the interests of the unit above their own. A sense of physical togetherness was induced, as each individual was immersed into the generalised whole.

The body was also at the heart of the experience of barrack life for recruits themselves. For some, it became prominent in times of failure, when men simply could not run fast enough or march in step. It seems that these men were constrained by their own physical limitations. Their bodies simply could not be moulded in accordance with the army's intentions, even when they tried to comply with the demands of their instructors. Instances where recruits experienced their bodies as a source of embarrassment or frustration therefore suggest the success of military power. Official bodily codes and ideals were internalised by recruits, who perceived their own abilities to cope with the rigours of training as central to their sense of self worth.[171]

Yet, while control and regulation were certainly extreme in these early months of service life, this does not mean that the army recruit was simply a subjected body, adapting to the demands of military discipline. Soldiers' testimonies highlight a range of tactics that men developed to counter the hardships and deprivations imposed by military duty. Some recruits consciously and openly directed their bodies towards opposition. This was most evident in cases of desertion, malingering and self-inflicted wounds, behaviours that were an obvious and outward response to the constraints of the military environment. Thus, men who did not want to go on parade went sick. Those who wished to evade the weekly cross-country run took the bus. Others, who wanted to go out at weekends or to go home for Christmas tried to sneak past the guards or simply absented themselves completely.

Resistance could also be discreet. During off-duty periods or in the safety of their barrack rooms, men were able to get drunk and have sex. These behaviours were not just about satisfying bodily impulses, but reflected wider ideas about status and masculinity. In the army, conceptions of manliness traditionally centred on the ability to drink to excess and to be promiscuous. This was a powerful influence on young recruits, who learned to drink beer or exaggerate their sexual experiences. Military staff also tolerated some behaviours that the army deemed unhealthy or unsafe. At times officers even conspired with their men, particularly in instances where they slept with each other. While publically adhering to official rules and regulations during working hours, in the evenings officers and recruits engaged in romantic relationships. These encounters allowed men to satisfy sexual impulses or even to assert their own control.

Men also experienced a sense of empowerment in instances where they complied with the army's designs for their bodies. Military service changed the attitudes and values of many newcomers, who came to enjoy the effects of training and confessed a sense of trust in the regime. Recruits became aware of their bodies not simply through their failures, but when they became fitter, stronger and healthier. No doubt, training moulded and improved bodies according to a concept of usefulness defined by the military agenda, but in turn, recruits engaged in their own physical transformations in order to make themselves look and feel good.

Training

Notes

1. French, *Raising Churchill's Army*, p. 127.
2. French, *Raising Churchill's Army*, p. 68.
3. War Office, *Basic and Battle Physical Training: Part I, General Principles of Basic and Battle Physical Training and Methods of Instruction* (London: HMSO, 1944), p. 9.
4. P. Grant, *A Highlander Goes to War: A Memoir, 1939–1946* (Edinburgh: Pentland Press, 1995), p. 8.
5. IWM SA, 17630, Eric Murray, reel 2.
6. MOA D 5165, Diary for November 1940.
7. IWM SA, 21565, Bill Partridge, reel 1.
8. MOA D 5134, Diary for December 1942, p. 2.
9. MOA TC29, Forces: Men in the Forces, 1939–1956, 2/A, Divisional Orders by Major-General P.J. Shears, Commander, Durham and North Riding County Division, 10 June 1941, p. 1.
10. IWM SA, 18435, William Dilworth, reel 2.
11. War Office, *Basic and Battle Physical Training: Part I*, pp. 21–2.
12. War Office, *Physical and Recreational Training* (London: HMSO, 1941), p. 4.
13. War Office, *Basic and Battle Physical Training: Part I*, pp. 26–7.
14. War Office, *Basic and Battle Physical Training: Part I*, p. 27.
15. See, for example, M. Mayhew, 'The 1930s nutrition controversy', *Journal of Contemporary History* 23 (1988), 445–56; Porter, *Health, Civilization and the State*, p. 186; Zweiniger-Bargielowska, 'Raising a society of good animals', pp. 77–8.
16. J. Vernon, *Hunger: A Modern History* (Cambridge, MA: University of Harvard Press, 2007), p. 87; Zweiniger-Bargielowska, *Managing the Body*, pp. 137–9.
17. Vernon, *Hunger*, pp. 96–104.
18. Long, *The Rise and Fall of the Healthy Factory*, pp. 31–2.
19. 'Quality and variety of army's meals', *Manchester Guardian* (17 December 1940), p. 6.
20. War Office, *Manual of Military Cooking and Dietary: Part 1, General* (London: HMSO, 1940), p. 6.
21. 'Quality and variety of army's meals', *Manchester Guardian* (17 December 1940), p. 6.
22. Crew, *The Army Medical Services: Administration, Volume I*, p. 376.
23. Colonel S. Lyle Cummins, 'Physical Development Centres', *Journal of the Royal Army Medical Corps* 18:3 (1943), 184.
24. Harrison, *Medicine and Victory*, p. 6.
25. In total there were three Physical Development Centres, the other two being opened in Skegness and Hereford. Crew, *The Army Medical Services: Administration, Volume I*, p. 380.

26 IWM SA, 19805, Walter Chalmers, reel 2.
27 IWM SA, 21565, Bill Partridge, reel 2.
28 IWM SA, 23195, Roy Bolton, reel 2.
29 IWM SA, 22387, Joe Stevens, reel 1.
30 MOA TC29, Forces: Men in the Forces 1939–1956, 2/D, Morale report 2, 15 December 1940, p. 5.
31 In 1941, the incidence of venereal disease among servicemen and male civilians had increased by 113 per cent. See M. Harrison, 'Sex and the citizen soldier: health, morals and discipline in the British army during the Second World War', in R. Cooter, M. Harrison and S. Sturdy (eds.), *Medicine and Modern Warfare* (Atlanta, GA: Rodopi, 1999), p. 227.
32 War Office, *The Soldier's Welfare: Notes for Officers* (London: HMSO, 1941), p. 12.
33 IWM SA, 18435, William Dilworth, reel 2.
34 IWM SA, 18512, Russell King, reel 5.
35 P. Ferris, *Sex and the British: A Twentieth-Century History* (London: Penguin, 1993), p. 43.
36 IWM SA, 30493, Ron Gray, reel 2.
37 MOA D 5061.1, Diary for April 1941, pp. 4–5.
38 MOA TC29, Forces: Men in the Forces 1939–1956, 29/E, Life in a depot – RAMC, p. 4.
39 'Masturbation in men', *British Medical Journal* (1 January 1944), 31.
40 War Office, *Manual of Military Law* (London: HMSO, 1940), p. 115.
41 Recent estimates suggest that 250,000 gay men served in the British armed forces between 1939 and 1945. E. Vickers, 'The good fellow: negotiation, remembrance and recollection – homosexuality in the British armed forces, 1939–1945', in D. Herzog (ed.), *Brutality and Desire: War and Sexuality in Europe's Twentieth Century* (Basingstoke: Palgrave, 2009), p. 115; P. Tatchell, 'When the army welcomed gays', 1996, www.petertatchell.net/military/when_the_army.htm (accessed November 2013).
42 IWM SA, 20891, Joseph Inskip, reel 1.
43 IWM SA, 23367, Henry Butterworth, reel 3.
44 IWM SA, 21565, Bill Partridge, reel 1.
45 See J. Terry, 'Anxious slippages between "us" and "them": a brief history of the scientific search for homosexual bodies', in J. Terry and J. Urla (eds.), *Deviant Bodies: Critical Perspectives on Difference in Science and Popular Culture* (Bloomington: Indiana University Press, 1995), p. 129; H. Ellis, *Studies in the Psychology of Sex, Volume 2: Sexual Inversion* (New York: Random House, 1937), p. 170. (*Sexual Inversion* was first published in Germany in 1896 and subsequently became the second part of Ellis's seven-volume *Studies in the Psychology of Sex*, originally published from 1897 to 1910); 'Anomaly', *The Invert and his Social Adjustment* (London: Bailliere, Tindahl and Cox, 1927), p. 9; C. Anderson, 'On certain conscious

and unconscious homosexual responses to warfare', *British Journal of Medical Psychology* 20:2 (1944), 162.
46 C.S. Jarvis, *The Male Body at War: American Masculinity During World War II* (Chicago: Northern Illinois University Press, 2004), p. 73.
47 C. Anderson, 'On certain conscious and unconscious homosexual responses to warfare', *British Journal of Medical Psychology* 20:2 (1944), 162.
48 IWM SA, 22075, Kenneth Bond, reel 4.
49 IWM SA, 17286, Douglas Arnold, reel 1.
50 IWM SA, 17628, Joseph Clark, reel 4.
51 IWM SA, 18435, William Dilworth, reel 2.
52 IWM SA, 21565, Bill Partridge, reel 1.
53 IWM SA, 18435, William Dilworth, reel 2.
54 IWM SA, 18512, Russell King, reel 5.
55 IWM SA, 17628, Joseph Clark, reel 4.
56 IWM SA, 18435, William Dilworth, reel 2.
57 IWM SA, 10601, Frederick Cottier, reel 4.
58 IWM SA, 20891, Joseph Inskip, reel 1.
59 IWM SA, 23216, William Corbould, reel 1.
60 IWM SA, 21565, Bill Partridge, reel 2.
61 IWM SA, 18512, Russell King, reel 5.
62 MOA TC29, Forces: Men in the Forces 1939–1956, 2/D, Morale report 2, 15 December. 1940, p. 2 and Morale report 3, 22 December 1940, p. 1.
63 MOA TC29, Forces: Men in the Forces 1939–1956, 2/E, Day to day life in the army, August 1940, p. 4.
64 Foucault, *Discipline and Punish*, p. 147.
65 IWM SA, 11468, Ian Sinclair, reel 3.
66 IWM SA, 13128, James Ford, reel 2.
67 MOA TC29, Forces: Men in the Forces 1939–1945, 2/E, J.A. Bergin, Blandford Camp, November 1940, p. 5.
68 IWM SA, 11468, Ian Sinclair, reel 3.
69 War Office, *Basic and Battle Physical Training: Part I*, p. 6.
70 War Office, *Basic and Battle Physical Training: Part II, Basic Physical Training Tables and Basic Physical Efficiency Tests* (London: HMSO, 1944), p. 19.
71 War Office, *Basic and Battle and Physical Training: Part II*, p. 15.
72 War Office, *Basic and Battle Physical Training, Part II*, p. 16.
73 Wellcome Archives, London, RAMC/1129, Handbook for Army Physical Training Corps instructors of No. 30 Physical Development Centre, February 1945, p. 26.
74 War Office, *Basic and Battle Physical Training: Part I*, p. 13.
75 IWM SA, 18435, William Dilworth, reel 2.
76 War Office, *Physical and Recreational Training*, pp. 6–7.

77 IWM SA, 1843, William Dilworth, reel 2.
78 War Office, *Basic and Battle Physical Training: Part I*, p. 9.
79 E. Durkheim, *The Elementary Forms of the Religious Life* (London: Allen and Unwin, 1976), pp. 230–1.
80 MOA TC29, Forces: Men in the Forces 1939–1956, 2/D, Morale report 2, 15 December 1940, p. 4.
81 Grant, *A Highlander Goes to War*, p. 11.
82 A.P. Wavell, *The Good Soldier* (London: Macmillan, 1948), p. 46.
83 IWM SA, 30493, Ron Gray, reel 2.
84 W.A. Elliot, *Esprit de Corps: A Scots Guards Officer on Active Service, 1943–1945* (Wimborne: Michael Russell, 1996), p. 105.
85 War Office, *Physical and Recreational Training*, p. 9.
86 J. Walvin, 'Symbols of moral superiority: slavery, sport and the changing world order, 1800–1950', in Mangan and Walvin (eds.), *Manliness and Morality*, p. 250; T. Mason and E. Riedi, *Sport and the Military: The British Armed Forces, 1880–1960* (Cambridge: Cambridge University Press, 2010), pp. 80–111.
87 Mason and Riedi, *Sport and the Military*, pp. 112–43, 178–216.
88 Wellcome Archives, London, RAMC/1129, Handbook for Army Physical Training Corps instructors of No. 30 Physical Development Centre, February 1945, pp. 31–5.
89 War Office, *Basic and Battle Physical Training: Part I*, p. 8.
90 IWM SA, 21565, Bill Partridge, reel 1.
91 IWM SA, 18512, Russell King, reel 5; IWM SA, 21565, Bill Partridge, reel 1.
92 IWM SA, 23216, William Corbould, reel 1.
93 Elliot, *Esprit de Corps*, p. 105.
94 IWM SA, 11468, Ian Sinclair, reel 3.
95 J. Hockey, 'No more heroes: masculinity in the infantry', in Higate (ed.), *Military Masculinities*, p. 16.
96 IWM SA, 20891, Joseph Inskip, reel 1.
97 MOA TC29, Forces: Men in the Forces 1939–1956, 2/D, Morale report 2, 15 December 1940, p. 4.
98 E. Goffman, *Stigma: Notes on the Management of Spoiled Identity* (Englewood Cliffs, NJ: Prentice-Hall, 1963), pp. 12–13.
99 War Office, *Basic Battle and Physical Training: Part I*, p. 29.
100 IWM SA, 21565, Bill Partridge, reel 1.
101 MOA TC29, Forces: Men in the Forces 1939–1956, 2/A, Letter from No. 1 Motor Battalion, 8 October 1940, pp. 1–2.
102 War Office, *Basic and Battle Physical Training: Part I*, p. 10.
103 Lt-Col R.A. Mansell, 'Man-management', *Journal of the Royal Army Medical Corps* 79:2 (1942), 72–3.
104 War Office, *Basic and Battle Physical Training: Part I*, p. 9.
105 IWM SA, 18512, Russell King, reel 4.

106 IWM SA, 20891, Joseph Inskip, reel 1.
107 IWM SA, 21565, Bill Partridge, reel 2.
108 IWM SA, 11468, Ian Sinclair, reel 3.
109 MOA TC29, Forces: Men in the Forces 1939–1956, 2/A, Report on morale and training in the army, 11 June 1941, p. 3.
110 IWM Document Archive, 2996, private papers of R.A. Graydon.
111 BBC WW2 People's War Archive, A3331577, Percy Bowpitt, 26 November 2004, www.bbc.co.uk/ww2peopleswar/stories/77/a3331577.shtml (accessed November 2013).
112 The official time period within which an absentee could demonstrate his intention to return was twenty-one days, after which he was classified as a deserter. See War Office, *Handbook of Military Law* (London: HMSO, 1943; first published 1929, reprinted 1939), p. 19.
113 IWM SA, 10601, Frederick Cottier, reel 4.
114 IWM SA, 18743, Robert Ellison, reel 3.
115 BBC WW2 People's War Archive, A331577, Bowpitt.
116 MOA TC29, Forces: Men in the Forces 1939–1956, 2/D, Moral report 1, 4 December 1940, p. 2.
117 MOA TC29, Forces: Men in the Forces 1939–1956, 2/D, Morale report 3, 22 December 1940, pp. 1–2.
118 MOA D 5061.1, Diary for January 1941, p. 6.
119 MOA FR 836, An army depot in 1941, August 1941, p. 6.
120 Lt-Col. W. Brockbank, 'The dyseptic soldier: a record of 931 consecutive cases', *Lancet* (10 January 1942), 41.
121 R. Good, 'Malingering', *British Medical Journal* (26 September 1942), 360–1.
122 R. Cooter, 'Malingering in modernity: psychological scripts and adversarial encounters during the First World War', in Cooter, Harrison and Sturdy (eds.), *Medicine and Modern Warfare*, p. 142.
123 S. Wessely, 'Malingering in historical perspectives', in P.W. Halligan, C. Bass and D. Oakley (eds.) *Malingering and Illness Deception* (Oxford: Oxford University press, 2003), p. 31.
124 M. Turner, 'Malingering', *British Journal of Psychiatry* 171 (1997), 409.
125 J. Collie, *Malingering and Feigned Sickness* (London: Edward Arnold, 1913); A. Bassett Jones, L.J. Llewellyn and W.M. Beaumont, , *Malingering: Or the Simulation of Disease* (London: William Heinemann, 1917).
126 Bourke, *Dismembering the Male*, pp. 85–92.
127 'Service medicine', *British Medical Journal* (23 December 1944), 834.
128 BBC WW2 People's War Archive, A2669222, Harold Pollins, 26 May 2004, www.bbc.co.uk/history/ww2peopleswar/stories/22/a2669222.shtml (accessed November 2013).
129 Good, 'Malingering', p. 362.

130 MOA TC29, Forces: Men in the Forces, 2/A: Reports from individual members of the forces, October 1940, p. 5.
131 R.C. L'E. Burges, 'Impressions of a regimental medical officer', *British Medical Journal* (6 December 1941), 816.
132 Major H.A. Sandiford, 'War neuroses', *Journal of the Royal Army Medical Corps* 71:4 (1938) 233.
133 J.A. Hadfield, 'War neurosis', *British Medical Journal* (28 February 1942), 284.
134 'Service medicine', *British Medical Journal* (11 November 1944), 643.
135 IWM SA, 16352, Frank Offiler, reel 4.
136 Good, 'Malingering', p. 360.
137 IWM SA, 10601, Frederick Cottier, reel 4.
138 IWM SA, 28681, Samuel Beard, reel 2 (used with permission from the BBC).
139 IWM SA, 16084, Edward Kirby, reel 5.
140 IWM SA, 18512, Russell King, reel 4.
141 MOA TC29, Forces: Men in the Forces 1939–1956, 2/E, J.A. Bergin, Blandford Camp, November 1940, p. 5.
142 MOA TC29, Forces: Men in the Forces 1939–1956, 2/D, Morris, Life in a depot – RAMC, January 1941, p. 4.
143 MOA TC29, Forces: Men in the Forces 1939–1956, 2/B, Leonard England, Morale report 1, June 1941, p. 3.
144 MOA FR Education in the armed services, May 1941, pp. 9–10.
145 For a discussion of similar activities in the US Army, see A. Bérubé, *Coming Out Under Fire: A History of Gay Men and Women in World War Two* (New York: Free Press, 2000), p. 37.
146 MOA D 5061.1, Diary for January 1941, p. 6.
147 MOA FR 686, Education in the armed forces, May 1941, p. 9.
148 IWM Document Archive 125, private papers of R.H. Lloyd-Jones.
149 MOA TC29, Forces: Men in the Forces 1939–1956, 2/E, Life in a depot – RAMC, p. 3.
150 See, for example, Bérubé, *Coming Out Under Fire*, pp. 45–6; M. Hirschfeld, *The Sexual History of the World War* (New York: Panurge Press, 1934), p. 124; W. Norwood East, 'The interpretation of some sexual offences', *Journal of Mental Science* 294 (1925), 414.
151 IWM SA, 17354, Dick Fiddament, reel 10.
152 IWM SA, 20318, Charlie Workman, reel 4.
153 See P. Fussell, *Wartime: Understanding and Behaviour in the Second World War* (Oxford: Oxford University Press, 1999), pp. 96–114; Woodward and Winter, *Sexing the Soldier*, pp. 70–4.
154 IWM SA, 23195, Roy Bolton, reel 3.
155 MOA D 5061.1, Diary for February 1941, p. 6.
156 MOA D 5165, Diary for November 1940, p. 4.

157 IWM SA, 17354, Dick Fiddament, reel 10.
158 IWM SA 30493, Ron Gray, reel 3.
159 MOA TC29, Forces: Men in the Forces 1939–1956, 2/D, Morale report 3, 22 December 1940, p. 2.
160 IWM SA, 10610 Frederick Cottier, reel 4.
161 BBC WW2 People's War Archive, A6006205, L.W.A. Lyons, 3 October 2005, www.bbc.co.uk/ww2peopleswar/stories/59/a6006359.shtml (accessed November 2013).
162 MOA TC29, Forces: Men in the Forces 1939–1956, 2/E, Life in a depot – RAMC, p. 4.
163 M. Houlbrook, *Queer London: Perils in the Sexual Metropolis, 1918–1957* (Chicago: University of Chicago Press, 2005), pp. 139–66.
164 M. Garber, *Vested Interests: Cross Dressing and Cultural Anxiety* (London: Routledge, 1992), pp. 44–56.
165 K. Clark and M. Holquist, *Mikhail Bakhtin* (Cambridge, MA: Harvard University Press, 1986), pp. 297–9.
166 Scott, *Domination and the Arts of Resistance*, p. 10.
167 Frank, 'For a sociology of the body', p. 42.
168 MOA TC29, Forces: Men in the Forces 1939–1956, 2/E, Aspects of army life, September 1940, p. 1.
169 MOA TC29, Forces: Men in the Forces 1939–1956, 2/E, Day to day life in the army, August 1940, p. 1.
170 IWM SA, 23195, Roy Bolton, reel 2.
171 B. Turner, *The Body and Society: Explorations in Social Theory* (London and Thousand Oaks: Sage, 2nd edn, 1996), p. 230.

3

Experimentation

Before the outbreak of hostilities in 1939 the MRC noted in its annual report that 'The history of war is, in one sense, largely a race between the development of instruments of physical destruction and the development of medical knowledge for saving the life of the sick and wounded.'[1] Indeed, the importance of scientific research to the military mission had been demonstrated during the First World War. Steve Sturdy refers to this as a 'grand experiment' in which 'the line of demarcation between battlefield and laboratory became increasingly blurred.' Whereas previously War Office officials had assumed that scientific research should be confined to periods between conflicts, as the war dragged on it became evident that victory would come only from the development of new equipment and strategies. As such, growing numbers of physiologists were inducted into war-related research, successfully applying their knowledge and investigative skills to develop new methods of fighting and killing the enemy, of protecting British soldiers from being killed, and new ways of treating the sick and wounded in order to return them to the fray.[2]

During the Second World War similar approaches were put into practice early on. At the start of the conflict armed service departments began working with researchers in the biological sciences in order to find solutions to the many and varied problems involving the 'human factor'.[3] As such, between 1939 and 1945 British soldiers were regularly engaged in a wide range of experiments that were designed to develop more effective weapons, treatments and modes of protection. These included trials of therapeutic drugs, synthetic stimulants and exposure to chemical agents.

Through an exploration of these various human trials, this chapter firstly examines how the soldier's body was perceived in discussions of experimentation between 1939 and 1945. It does so by looking at the three main objectives behind wartime human research, which can broadly be defined as: enhancing the body, harming the body and restoring the body. Cutting across the first two was a conception of the healthy body as both limited and limiting, as military researchers strove to widen the parameters of physical performance. Within the third agenda lay a conception of bodily usefulness based on the production of medical knowledge, as the war-wounded bodies of soldiers provided civilian scientists with a valuable experimental base on which to apply their theories and techniques. The second objective of this chapter is to answer the question of why men participated in human trials. Many existing studies of human experimentation within the military have framed participation within wider debates about formal ethical frameworks. These have often emphasised the coercive nature of the military as an institution.[4] This chapter suggests that in order to understand why men took part in experiments during the Second World War, it is necessary to widen the scope of analysis. Soldiers' testimonies allow us to begin reconstructing the agendas and decisions of those who participated and show that a wide range of factors influenced how and why men found their bodies subject to experimentation.

Enhancing the body

The soldier's body was, in the first instance, something to be enhanced, as researchers strove to determine the optimums and limits of physical capacity. In 1940 the MRC set up the Military Personnel Research Committee to devise 'ways and means of ensuring the maximum safety, efficiency and comfort of the healthy soldier on active service'.[5] Staffed by a combination of civilian scientists and military personnel, its work comprised a range of experiments that were designed to improve the physical performance of troops in different operational settings. These included studies on types of clothing, body protection, tolerance to climatic extremes, particular hazards and susceptibility to certain illnesses and conditions.[6] In 1941, for example, a team working at the National Institute for Medical Research in Hampstead carried out a series of trials to test human susceptibility to motion sickness, which was a particular threat to the efficiency of air and seaborne troops. These required soldiers to swallow a balloon attached to a recording device before being placed on a specially designed swing. They were then subjected to violent movements and loud noises while their stomach contractions were counted. As a result of these tests the researchers were able to ascertain that individuals with an absence of aural labyrinths (the part of the inner ear attuned to gravity and motion) were insusceptible to the condition.[7]

In 1942 a series of 'hot room' experiments were conducted on army personnel at the London School of Hygiene in order to determine the limits of human tolerance to heat. These were designed to enhance the performance of tank crews serving in Africa and the Far East. By successively exposing the participants to 'tropical conditions' in the rooms, the researchers ascertained that 'men are apparently physically efficient and able to continue for periods of up to two hours with rectal temperatures of 102°F. (38.9°C.). But, as a result of continued observation and personal experience, a limit of 101°F. (38°C.) has been taken as an acceptable upper limit.'[8] The researchers also determined that the human body could become acclimatised to the effects of heat as with each exposure the men sweated more and their temperature did not rise so high.[9] In addition, a Dr McArdle was able to design a ventilated belt which brought outside air straight under the clothes of men in tanks, thus enabling them 'to carry on in conditions which would otherwise be physiologically intolerable'.[10]

Military researchers also sought to try and enhance the body's natural powers of endurance by experimenting with amphetamines. Reports from enemy sources in the early stages of the war had revealed that analeptics were being used by German tank crews and other troops in North Africa. Following this, the MRC set up a Sub-Committee on Analeptic Substances in 1941 in order to conduct human trials with the stimulant drug Benzedrine.[11] The first was carried out in early 1942 on men from the 7th Canadian Infantry Brigade. Some of the subjects were given a 'special emergency ration' of 5 mg Benzedrine sulphate in chocolate. Others were given a 'dummy emergency ration' of chocolate alone, and a final group were given nothing at all. Throughout the experiment the men were all subjected to various training tests and tactical exercises.[12] As a result of these trials it was concluded that neither a single 5-mg dose nor a 5-mg dose repeated once caused any appreciable improvement in the capacity of the troops to carry out long marches, to dig themselves in or to use weapons, even though moderately severe fatigue was induced by the nature of the exercise.[13]

In June 1942, therefore, further trials were carried out on 'severely fatigued' men by Dr D.P. Cuthbertson and Dr J.A.C. Knox from the Institute of Physiology at the University of Glasgow. Six subjects drawn from the RAMC and Royal Army Service Corps were kept awake for twenty-four hours, performing light drills, games and gymnastics. They were then treated with either 10–15 mg of Benzedrine or 10 mg of Methedrine, or again with a 'dummy tablet' containing no analeptics.[14] The experiments showed that both Benzedrine and Methedrine caused increased wakefulness in fatigued men and that 'the subject's capacity to sustain a given level of work performance was increased within ½ to 1½ hours and was sustained for about 1 hour.'[15] Although 'wide individual variations' in responses to the drugs were observed, it was decided that these substances might be useful in helping fatigued troops whose task was likely to last less than five hours, or whose task was of such a character that hope of survival was negligible except by supreme effort. Finally, a set of field trials was organised on armoured troops serving in the Middle East. Thus, the battlefield also became the laboratory. Observers concluded that Benzedrine was of significant value in maintaining the efficiency of men who had been awake for up to forty-eight hours.[16] In doses not exceeding 20 mg per twenty-four hours, it enabled 'a moderately fatigued man to work better and a severely fatigued man to work longer...the effects began after about an hour and began to decline about 5, with appreciable effects for about 8 hours after administration.'[17] These experiments therefore highlight a conception of the body as limited in its natural state but also malleable, with the potential for manipulation and improvement towards optimums of physical performance.

In other experiments the body was perceived as limiting, particularly in relation to the machinery that it operated. The MRC noted that 'In many cases the main limiting factor in the efficiency of the machine or weapon is the human being working it. The earlier the stage at which the human factor is considered by physiologists and psychologists in relation to machine or instrument design, the more effective the end result.'[18] As such, a range of studies emerged during the war which tried to harmonise man and machine. In 1940 a Physiological Research Laboratory was established in the Gunnery Wing of the Armoured Fighting Vehicle Training School at Lulworth to investigate the physiological factors that interfered with efficient tank-driving.[19] This included a series of tests on the seat-to-pedal relationship, the design of steering sticks and the relationship of various levers to the seat. In 1943 researchers also surveyed the body measurements of several Royal Armoured Corps personnel. By applying this data to tank design they were able to achieve a proper utilisation of 'crew space'.[20]

This type of research drew largely on expertise developed in the industrial world and again shows a connection between approaches to the soldier's body and the civilian worker. During the interwar years the Industrial Fatigue Research Board had carried out experiments on 'the design of machinery in relation to the operator' which were based on the notion that the efficient functioning of most machines depended 'to a greater or lesser extent on their operator' and should be designed 'to meet his physiological requirements'.[21] Thus, during the war, members of the Industrial Health Research Board became involved in military research. Secretary R.S.F. Schilling noted in 1944 that 'since the outbreak of war the Board's investigators had given up much of their time to carrying out research and giving advice on non-industrial problems'.[22] As such, at the end of hostilities, the methods and techniques developed in the sphere of military research could be reapplied to industry:

> It is already clear that, just as the methods and results of industrial health research have assisted in the solution of various Service problems, so also many of the research methods adopted to solve urgent problems affecting the Fighting Services under war conditions can be effectively applied in peacetime to the solution of numerous problems of the human factor in industry; these include not only the selection of the right worker for the right job, and the securing of the best practicable environment for the job, but also the design of machines, instrument panels, working benches from the point of view of maximum comfort and efficiency.[23]

A reciprocal relationship seems to have been established, as the soldier's body became synonymous with that of the industrial worker. Whether on the battlefield or factory floor, the body was a resource, something to be made efficient in its designated role.

Harming the body

In the quest for military victory the soldier's body was also something to be deliberately harmed through its participation in experiments. This was chiefly the outcome of exposure to weapons so that scientists could observe their physiological effects. This, in turn, would enable researchers to develop more effective ways of disabling the enemy and of protecting British troops in the field. In 1941 the Military Personnel Research Committee's Weapons and Biological Assay Sub-Committee conducted investigations to decide the effects of 'blast' from a variety of weapons upon personnel in confined spaces such as pill-boxes, armoured fighting vehicles and buildings converted for defensive purposes.[24] In the same year a Sub-Committee on Body Armour and Steel Helmets conducted field trials in North Africa to test the protective value of a new type of

Experimentation

body armour against various bullets and weapons.²⁵ Most of this type of research on the soldier's body was, however, directed towards chemical warfare. Between 1939 and 1945 over 7,000 servicemen took part in 'special tests' in which they were deliberately exposed to toxic substances at the Chemical Defence Experimental Establishment at Porton Down in Wiltshire.²⁶ The station had been established in 1916 by the Ministry of Munitions and was staffed by servicemen, civilian and technical civil servants, and other civilian workers. Originally it had confined its research to animals, with occasional tests on humans during the First World War. However, since 1922, soldiers had regularly been employed as research subjects, particularly to develop knowledge of mustard gas.²⁷ Most of the men who attended came from the army's lower ranks, such as privates and sappers. The majority were drawn from the infantry or the non-combat Pioneer Corps. These were soldiers who were considered to be in the lowest intelligence groups and who mainly carried out labouring, supply and administrative duties.²⁸

In the early years of the war the research was mainly offensive as scientists focused on developing new chemical agents, complementing existing ones and designing new weapons in which they could be delivered. Thus, the body was again pushed to the limits of endurance so that scientists could ascertain the required levels of chemicals necessary to 'break' a man.²⁹ Gas substances such as mustard gas vapour, lachrymators (tear gases) and sternutators (sneezing agents) were dispersed atmospherically in chambers and the onset and severity of men's symptoms noted. Liquid compounds, on the other hand, would be placed directly on the skin or in the eyes.³⁰ Soldiers also took part in field trials so that scientists could assess the effects of different agents and weapons on men working under more realistic battle conditions. In January 1942, for example, a group of eleven soldiers were fired on with mustard gas explosives while digging, manning a Bren gun and preparing to shoot weapons. All of the men sustained burns considered to be of 'casualty-severity', most being unfit for duty for thirty-two days.³¹ As the war continued, however, the priorities for research at Porton changed as the authorities came to realise that new agents could not be developed in time to affect the war in Europe. Scientists therefore turned their attention to developing effective countermeasures against agents that the enemy might deploy. This included studies on the design of respirators (gas masks) and the protection offered by different types of clothing. For example, subjects had small patches of contaminated material fixed to their arms or wore articles of impregnated clothing over long periods of time so that scientists could assess their protective qualities.[32] Experiments on decontamination also involved the body first being contaminated before a treatment such as an ointment, cream, cake or powder was applied and its effects observed.[33] Research subject Stanley Shore, a member of the 112 Royal Armoured Corps, who took part in an experiment involving the decontamination of foodstuffs laced with lewisite, recalled that 'They contaminated the food and then decontaminated it, but you didn't have to eat it. They sort of strapped it on your arms in little containers and left it on there for two or three days to see if it blistered. If it blistered of course they knew that their decontamination wasn't up to whatever they were trying.'[34] As such, the damage caused to the soldier's body depended on the efficacy of the remedy being tried out.

The ways in which bodies were assessed within these experiments were related directly to their military capacities and not necessarily to the health or wellbeing of the individual. Reporting on a trial to determine the casualty-producing rate of airburst mustard gas shells, Porton researcher Captain Curwen noted, 'In selecting casualties, the men chosen were those whom it was considered under actual service conditions would be unable to carry out any useful military task.'[35] This ethos was also evident during a trial to determine the disability produced by exposure of the skin to mustard gas vapour. In his report, Dr Sinclair from the Department of Anatomy of Oxford, who worked at Porton during the war, explained that 'the term "disability" can be accurately defined and graded in a manner which bears some relation to the requirements of the military situation.' Elaborating further, he stated that 'So long as disabled men retained the capacity to fire a weapon actively they continued to be classified as "partially disabled" even when they were sent to hospital on account of their injuries. The justification of this procedure is that, from a military standpoint, such men, whom it might not always be able to evacuate, could be of considerable value to the defence of a fixed position.'[36] The body was further de-individualised as each was evaluated according to 'clinical signs and data' and not the experience of the participant himself.[37] It is significant to note that subjects at Porton were referred to as 'observers' because, according to the official histories, they 'observed the effects of chemical warfare compounds on themselves and described those effects to scientists'.[38] Yet experimental reports reveal that this often did not occur. In trials to assess the effect of lachrymators on vision, for example, N.H. Mackworth, a member of the MRC, noted that the 'Curwen-Somerville technique' was adopted. This meant that 'all facts on the clinical appearance of the subjects were kept strictly apart from data on bodily sensations.' He explained that 'it is based on the number of clinical signs and symptoms present. It avoids the dangerously misleading practice of asking the subject of his ability to aim and fire a rifle.'[39] Indeed, when one in every ten subjects claimed to have experienced slight mental confusion and fright, Mackworth suggested that this was 'not possible' as 'the conditions were quite unfavourable for the development of such general psychological changes.'[40]

Restoring the body

It was not only the healthy military body on active service that became the focus of scientific inquiry during the Second World War. The war-damaged body was also the object of experimentation as researchers sought to develop more effective methods of treatments in order to facilitate the quick return of men to active duty. Reporting on the medical aspects of the war in 1946, W.E. Le Gros Clark from the Department of Anatomy at the University of Oxford stated, 'in every war, research workers tend to turn their attention once more to a consideration of the fundamental processes of the repair of damaged tissues and the restoration of disturbed function.'[41] This endeavour can clearly be seen in the work of the MRC's War Wounds Committee. Established in 1940, its work included research into blood transfusion, traumatic 'shock', war wounds, burns, brain injuries and nerve injuries.[42] Through its studies on blood transfusion, for example, soldiers' bodies were reconstituted with parts of other human bodies. The Transfusion Research Committee and Transfusion Service conducted research on blood substitutes such as plasma and serum, which were found to have a longer shelf life than whole blood.[43] The Committee also experimented with different types of dressings and plasters in order to repair damaged limbs. In 1940 soldiers took part in large-scale clinical trials of a bag made from carbon-impregnated cloth which enclosed limb and plaster and was designed to eliminate the smell emanating from wound discharge in normal 'closed plaster' treatment.[44] The Committee's work also took investigators to combat theatres in order to access relevant scientific material. In 1943 British Traumatic Shock Team No. 1, a group of RAMC workers, was deployed to the forward area in Italy to study the effects of severe limb injury. This was followed by British Traumatic Shock Team No. 2 in 1945, which worked in north-west Europe gathering blood and muscle specimens from almost 200 wounded soldiers.[45] By travelling to theatres of combat, investigators therefore got the chance to apply their knowledge and skills to those bodies that were likely to generate good scientific results. Highlighting the benefits of working on the front line, in April 1945 Lieutenant-Colonel R.T. Grant, the officer commanding Traumatic Shock Team No. 1, highlighted the benefits of working on the front line when he stated that 'the conditions of warfare in Italy provided an opportunity for the study of the injured beyond our expectation. There was no lack of casualties and we were given every facility for access to them.'[46]

Perhaps the most significant research programme in this respect was with the therapeutic drug penicillin. The historian Peter Neushel suggests that 'the penicillin programme was one of the largest wartime initiatives and is undoubtedly among the most successful research and development ventures ever.'[47] The drug had been discovered by Professor Alexander Fleming in 1929, was developed by Sir Howard Florey and Dr Ernst Chain in Oxford from 1939 and was shown to prevent infection of septic wounds. As such, a small quantity was despatched to the Middle East in April 1943 for experimental use on wounded soldiers. Then in May 1943 special investigators, including Florey, were sent to conduct experiments on troops suffering from chronic septis in North Africa. As a result of this work they ascertained that 'it was far too late to start penicillin treatment weeks or months after wounding, at a Rear Base hospital, and that its use should be tried much earlier, before the establishment of serious infection.'[48]

Penicillin research certainly privileged the military body over that of the civilian, as soldiers gained access to a therapy that was not made available to the rest of the population. This was due to limited supplies of the drug and the difficulties in producing it on a large scale when the country's resources were being directed elsewhere.[49] Thus, in the early stages of clinical trials it was decided that first priority should be given to the treatment of patients with war wounds.[50] Indeed, when four research centres were set up in hospitals in Oxford and London, their locations had to be kept a secret due to the great flow of requests for from the public to be included in the trials.[51] The main objective behind these wartime experiments was, however, not to cure the individual soldier, but rather to use his body to generate new medical knowledge. This again was exacerbated by the limited supplies of vital therapeutic materials, which meant that, even within the military, bodies were carefully selected for inclusion in particular trials. Basil Reeve, a doctor who worked with Traumatic Shock Team No. 1 in Naples, recalled in an interview that trials with blood transfusion were not conducted on men with severe pelvic wounds because 'we could keep them alive, but, you know, we would use up masses of blood.'[52] In an article in the *British Medical Journal* the Penicillin Clinical Trials Research Committee also made explicit its intentions 'to gain new knowledge of the curative possibilities of the drug rather than merely to repeat the therapeutic successes of which it is already known'.[53] Thus, owing to the short supplies of the drug, experimentation within the military had to be confined only to those bodies that would further knowledge of the mode of using the therapy. These bodies were identified as those that would help 'define the minimum effective dosage, the best methods of administration, and any factors not yet studied on which success may depend, and to explore the possibilities of penicillin treatment in conditions hitherto unstudied from this point of view'.[54] While the MRC claimed that 'This was a difficult objective to reconcile with humanitarian claims,' it nevertheless refused treatment which might have proved beneficial to individual patients but would not advance knowledge of the drug's action.[55]

Indeed, it appears that to the researchers the military nature of the body could be insignificant or incidental, as warfare had simply provided a unique opportunity to develop their theories and techniques. The push for victory had created a necessity to develop new therapeutic practices, while the bodies of wounded soldiers provided a valuable experimental base on which to apply medical knowledge and skills. Emphasising the value of war for wider scientific research, Traumatic Shock Team No. 1 worker Basil Reeve explained that 'to make advances you really need enough patients...you only see occasional patients, but in the war you see many patients and you're presented with the urgency of the problem... It's very difficult to find in ordinary civilian life, situations like the battles and the battle casualties you see...if there was a really bad epidemic, say of some viral disease, that would be interesting to study.'[56]

Certainly, the MRC asserted that although its wartime experiments with penicillin were conducted in a military context, they were simply an extension of existing scientific research. It noted that 'Statements have appeared from time to time that the work on penicillin at Oxford was started as an attempt to contribute to the treatment of septic wounds in the Second World War. These give a false impression, as the work, of which wound treatment later became a part was initiated well before the outbreak of hostilities.'[57] This opinion was later echoed by J.W.S. Pringle from the Department of Zoology at the University of Oxford in a statement to the Royal Society of London. He claimed, 'it is incorrect to suppose that work on penicillin was started in attempt to control septic wounds; the real situation was that an academic study with possibilities of wide theoretical interest was ready to be adapted to the needs of war.'[58] Discourses of legitimisation for the army therefore conceived of the body in terms of its potential to further the aims of medical science in the case of therapeutic trials. Neither the wellbeing of the individual body nor its military efficiency was regarded as a core objective in this context.

A moral economy of bodies

While it is clear that the soldier's body in wartime was a useful experimental object that allowed the authorities to develop more effective weapons, treatments and modes of protection, the question remains of how these experiments were organised and more specifically why men participated in them. On the one hand, it appears that by entering the military the body was expected to be placed at risk, be it on the battlefield or in the laboratory. At Porton Down chemical research was conducted almost exclusively on service personnel, although in 1926 approval was given for civilian laboratory assistants to volunteer.[59] In July 1944, in order to meet the demands of the war, the more general issue of civilian volunteers was considered, but was turned down because of 'complicated questions of compensation, remuneration, etc.'.[60] For the soldier, however, it had been decided in 1930 that any injuries sustained as a result of experimentation were 'attributable to the service' and paid from army funds as the Ordinary Regulations provided. When conflict broke out in 1939 the War Office also resolved that if a subject was injured at Porton, he would be paid from the Ministry of Pensions, in the same way as if he had been wounded in battle.[61]

Yet it was not only soldiers' bodies that were being used in experiments between 1939 and 1945, even for research that was directly war related. In 1941 Kenneth Mellanby, Research Fellow at the Sorby Institute at the University of Sheffield, conducted a series of experiments on a group of conscientious objectors on the transmission of scabies – something that had become a serious menace to the health of the army. The participants slept in between sheets that had been infected by soldiers or wore contaminated underclothing continuously day and night for a period of at least a week. Mellanby's rationale for choosing conscientious objectors was that 'they were the only section of the population not likely to be called up or compelled to leave the experiment due to military or industrial necessity'.[62] In 1943, an Army Malaria Research Unit also conducted trials in Oxford with the prophylactic mepacrine. The drug was administered over long periods of time to 200 research subjects, including students as well as men from army units.[63] Likewise, the MRC set up a Jaundice Committee in 1943 to study the mode of transmission of what had become a serious drain on manpower among soldiers serving in the Mediterranean. These experiments required humans to be infected with hepatitis. Finding participants was particularly difficult as it had a long incubation and illness period. The team again chose a small number of conscientious objectors and 350 individuals with rheumatoid arthritis, based on the knowledge that its symptoms could be relieved by an attack of jaundice.[64] In a context of warfare, therefore, it seems that the military body could be prized as something not to be risked in experimentation, as the demands of manpower meant that scientists looked to other groups to conduct their human research.

The experiments conducted on soldiers' bodies during this period also appear to have taken place within an informal ethical framework, or moral economy, which guided both the actions of researchers and their experimental participants.[65] In order to understand why this was the case the experiments must be placed within a wider historical context. As Jordan Goodman, Anthony McElligot and Lara Marks have argued, although the Nuremburg Code of 1946 may have institutionalised the issue of informed consent by setting down a formal list of regulations, it should not be seen as the beginning of ethical human experimentation. They suggest that prior to this many experimenters had also been regulated by informal, socially sanctioned codes of behaviour.[66] Indeed, it appears that in early twentieth-century Britain human experimentation was conceived within a discourse of morality, if only to safeguard the state against financial losses. In 1933 the Treasury advised the MRC that the informed consent of participants involved in influenza research was necessary in order to avoid any later claims for damages. It recommended that the nature of the risk be explained to all participants in experiments and that researchers conduct themselves with 'all due care and take all precautions suggested by medical science'.[67] These moral considerations were certainly evident within the wartime army in experimental contexts where there was the potential for bodily harm. At Porton Down, all three of the armed services had been supplying experimental participants since 1929. By 1938 sixteen men per week were required by the station and by then each of the services were supplying subjects in rotation. In June 1939, however, this system was not deemed satisfactory and the Chemical Defence Research Department suggested that each service should be requested to provide eight subjects per week. The three services therefore put their various commands on monthly rosters. Nevertheless, almost immediately the navy and the air force ceased to supply volunteers and the full responsibility fell on the army until the navy re-joined the scheme late in 1944.[68]

The method for recruiting servicemen was referred to as the 'observer scheme'. The station informed the War Office of the types of tests and number of volunteers needed. This information was passed from the War Office to the service ministries and on to service units, which then called for volunteers with information about the tests and administrative arrangements. The call was often made in written form, called the 'recruitment notice', or men were sometimes verbally invited to participate during parades.[69] Early on in the operation of this system concern was expressed at Porton that some of the subjects who arrived 'were not fully aware of the conditions under which service personnel were asked to volunteer'.[70] At the station's request, therefore, the War Office wrote to service departments emphasising that it should be explained that participants would be required to submit to the effects of gas, defined as 'certain liquids that may blister the skin when applied to it'.[71] Staff at Porton also suggested the form of words that might be used in recruitment notices. In turn, the ministries advised units on how the notices should be phrased. In 1940 a memorandum to the Army Council from the War Committee regarding the exposure of men to sternutators advised that notices should read, 'The tests involve exposure to toxic gases and liability to skin burns, and are subject to conditions and safeguards approved from time to time by the Army Council. All tests are carried out with every possible care, and under the direct supervision of medical officers, whose duty it is to ensure that those who do volunteer will incur no danger to their health.'[72] Clearly, there was a tacit recognition that men sent for experimentation by the armed services should be aware of what they were letting themselves in for. Indeed, a report from Porton went further, and stipulated explicitly that 'no form of moral or disciplinary pressure is brought to bear to persuade an individual to partake in a test against his personal wishes.'[73] Participation was to be informed, and it was to be consensual.

It is, however, important to note that at each stage the forms of words were suggested rather than prescribed. It was not until 1964 that a uniform procedure was put in place so that notices calling for volunteers appeared in the official administrative instructions regularly issued by the services and the Ministry of Defence. It was the 1980s

Experimentation

before signed consent forms were introduced.[74] As such, it is not always clear if what soldiers were told was what had been devised by those organising experiments or if it was reworded by those passing on the message. This, after all, had to work through layers of bureaucracy and military command before men heard it. Soldiers' testimonies are therefore crucial in providing evidence of what was actually announced and what information was available about the research to which they subjected themselves. The task of uncovering men's experiences of experimentation is inherently problematic due to the swathes of official secrecy that have historically surrounded military research.[75] Only a few of the oral history recordings in the Imperial War Museums Sound Archive contain the testimonies of soldiers who participated in wartime experiments. However, in 1999 Wiltshire Police launched an investigation into allegations of misconduct at Porton Down, during which it was in contact with around 600 servicemen who had attended the station throughout its history. As part of a historical survey of Porton in 2001 the Ministry of Defence invited these men to respond to a questionnaire about their visits.[76] These sources therefore can suggest the reasons why men took part in human trials.

Some soldiers have clearly articulated that they were coerced into taking part in human trials during the war. As we have seen, military training had been designed to condition men to be ordered and compliant and to follow orders without question. Thus, when recalling how he ended up at Porton Down, one army veteran stated, 'I was a raw recruit and did as I was told. No notice about risks etc. just detailed to go being just called up for service in June 1940'.[77] Another suggested that he had been misled into thinking that it was part of his military training. He reported that 'I was ordered. I understood that I was going on a gas course with no experiments'.[78] In an interview, Edward Kirby, a recruit in the Royal Engineers, could remember being put into an army lorry and being taken along with other men in his unit to Porton Down. There they were exposed to the sneezing gas DM (Adamsite):

> The DM gas, which we had first-hand experience of because we were asked to go to Porton Down, not asked but told [laughs] to go to Porton Down. We were taken by lorry and we were put in a room and they released with DM gas in this room and the effects were really terrible because they induced headaches and vomiting and complete incapacity for any kind of physical activity so you can imagine an enemy being subjected to this gas would be quite incapable of fighting. It wasn't lethal but it was quite debilitating so that you spent the whole of that day being sick.[79]

Wilfred Hall similarly recalled being ordered by his lieutenant to attend Porton Down while his unit was on parade in 1940. He also expressed a real fear of repercussion if one did not comply:

> I suppose being young you thought it was your duty to do it. While you was in the army, like I say, they told you to do these things and you should obey the last word of command, which [laughs] I know you should do, but sometimes you don't...I suppose in a way you could...you could either have been discharged out the army. That's the thing. I wouldn't even know what they'd do. I mean it happened in the 1914 war, didn't it, when they shot those people? So whether it was the fact of, well, would they have done anything like that, which I think you don't really know.[80]

Other men have claimed that the nature of the trials was not fully explained to them, or that the nature of the risks was misrepresented to them. In total there were thirty respondents to the Ministry of Defence's historical survey who had participated in experiments during the 1940s.[81] Almost half of these men claimed that no information was given to them about the substances to be used or whether they might feel any discomfort. Of those who stated that they were recruited by notices, 29 per cent remembered the notice saying something about the purpose of the trial and none remembered it saying anything about the risks that the trial might involve.[82] Military command culture therefore seems to have been responsible for the provision of recruits for chemical experiments in the Second World War. While some men stated that they were simply obeying orders and that no effort was made to seek their consent, others have suggested that little consideration was given to informing them about the experiments to which they subjected their bodies.

However, the evidence is more complex. Soldiers' testimonies suggest that many were active volunteers who willingly engaged their bodies in the experimental process. One wartime respondent to the Ministry of Defence's questionnaire commented that 'an officer came to our unit and asked for volunteers,' while another stated, 'I was curious so volunteered.'[83] Although it is not clear what information these men received, it does seem as if they felt that they had given their consent. There were various reasons why soldiers made the choice to participate in experiments. Rob Evans, a journalist who has researched Porton Down, argues that most men did volunteer and thought it was just part of their service to the nation.[84] Other stories suggest that participating in experiments allowed men to adopt new roles or to enhance their perceptions of themselves.

Experimentation

After the Military Personnel Research Committee's swing-sickness trials, for example, a group of seventy men were taken out to sea in rough weather in order to test the efficacy of various drugs. Researcher H.E. Holling noted that, 'since some of the men embarked with an air of secrecy carrying martial weapons they were sometimes mistaken by onlookers for a commando raiding party, a mistake which pleased the men.'[85]

At Porton Down experiments also became tests of character where the body's ability to endure became a badge of masculinity. As we have seen, the station predominantly employed low-ranking, non-combat recruits – men who had a limited outlet for performing the traditional man-of-action ideal. As such, by participating in experiments, these men could perhaps reassert a sense of status, if only momentarily. Reporting on a preliminary set of chamber trials, Mackworth reported that 'Their normal interest in the external world had faded and the one thing they wanted was an end to their very unpleasant bodily sensations. To obtain this, 9 out of 25 subjects who were exposed to the vomiting agent diphenylcyanoarsine (DC) were willing to risk subsequent derision from the rest of the men by leaving the chamber before completing the test.'[86] Indeed, this sense that experiments were tests of masculinity and capacity was played upon by researchers in order to generate good scientific results. When it came to a set of field trials in which soldiers were contaminated with arsenical smokes while completing an assault course, the men were divided into teams and each timed separately. As Mackworth explained:

> To encourage really competitive spirit the men were all told these times as soon as they had finished. After they had rested for 2–3 minutes they were sent to a point at the side of the course about 50 yards from the wide ditch so that they could shout and urge on the other subjects. During the last three days of the experiment the company commanding officer or the company sergeant major was also stationed at this point for the same reason.[87]

These examples therefore highlight a range of factors that may have affected why men subjected their bodies to experiments and indeed how they behaved within the tests. Their bodies were central to the ways in which they perceived themselves and to the opportunities that experiments provided to enhance their self-perception, be it in terms of their service to the nation or their own competencies and masculinities.

There were also more practical advantages to be gained from participation in human trials. Holling, for instance, highlighted the pleasures of a day out when he explained that 'the majority of men enjoyed the trips, even though they might be sick, for a sea trip was a pleasant change from the intensive training they were undertaking at the time.'[88] As part of his scabies experiments in Sheffield, Mellanby also conducted a series of therapeutic trials on infected soldiers. He suggested that 'The soldier patients showed no objection to being admitted to the institute; in fact they seemed to enjoy their stay! They did at least get a proper bed to sleep in, with real sheets; this was greatly appreciated, and for the first 24 hours most of them were sound asleep when they were not being treated.'[89]

Inducements could be more material still. At Porton Down, for every ten completed gas-chamber tests a soldier would receive seven days additional leave. All subjects there were free to spend their time as they wished when they were not taking part in experiments. A bus service even ran into Salisbury every evening.[90] For Private Stanley Shore, these were incentives enough to persuade him to volunteer. He did so because the station was close to his parents' house and he was able to go home and visit. He also recalled the better food and opportunities for nights out when he stated, 'The food was excellent. We even had oranges on one occasion. I don't know where they came from. We'd never seen one of those for a long time. We had transport into Salisbury at night if you wanted time off. They always used to check you over in the morning to make sure you hadn't got a hangover or anything like that, just to see if you were all right, you know, fit for experiments.'[91]

The body was most clearly commodified both by the authorities and the men themselves for the purposes where it was exchanged for cash. Early in 1945 Ralph Kirkton volunteered to take part in experiments with amphibious tanks because 'it added two and six to my pay packet and two and six in those days was an awful lot.'[92] In 1924 the Treasury had also sanctioned additional pay for volunteers participating in tests at Porton Down. Service personnel received one shilling for every physiological test, sixpence for every breathing test and sixpence for any other tests.[93] This could be a substantial addition to the wages of a private soldier, whose typical pay was twenty-one shillings per week.[94] Certainly, Stanley Shore felt that experimentation was worth it. He explained that 'you got possibly a shilling or so every time you went into the gas chamber. I mean, I couldn't be sure of that, but of course a shilling was a lot of money in those days.'[95] Additional payments were also made for each injury incurred. Edward Kirby explained that 'We were given a shilling. One shilling and that was enough to tempt you, so that was that. We were

Experimentation

submitted to mustard gas and you got half a crown a blister, for all the blisters you got, you got half a crown for every blister because the mustard gas affected your skin.'[96] Experimentation was therefore another site of negotiation between the individual and the state. In exchange for money or time off, soldiers chose to engage their bodies with the demands of military and medical science.

Researchers were not always successful in imposing their agendas on the bodies that they employed. Harold Pollins became involved in penicillin trials after acquiring septicaemia from boils on his legs during basic training in 1943:

> In due course I found myself at a military hospital in Penshurst. The boils and blotches on my legs appeared to the medical staff to be suitable scientific specimens for a research project. My medication consisted of one leg being treated in the traditional way with cloths made wet by being immersed in, I think, gentian violet (at least it made everything they touched that colour). The other leg was treated with penicillin and left uncovered...The research project was I'm sure a failure. I was not a very good subject. I could not help scratching the leg which had been treated with penicillin and the medical officer was greatly annoyed with me for ruining the research. In a camp sort of way he flapped his hand at me and said, 'naughty, naughty boy' or something to that effect.[97]

While Harold's imperfections as an experimental subject were therefore admonished by the military command structure, this was done in an ineffective and almost ironic way, by the doctor 'flapping his hands' and calling him a 'naughty, naughty boy'. In the end, Harold's own physical discomfort led him to frustrate the experiment and perhaps even his own desire to comply. Ultimately, he wanted to scratch his leg. Indeed, he stated that he 'could not stop scratching'. As such, it seems that his body was beyond the immediate control of both the doctor and Harold himself.

Other men intentionally resisted the experimental process. For instance, although the minesweepers in Holling's sea-sickness trials were volunteers and enjoyed their trips out to sea, some seemed reluctant to take the pills which could have unpleasant side effects. These men did not swallow their tablets but spat them back out into their drinking cups in an attempt to fool the observers.[98] In March 1942 the Military Personnel Research Committee's Rations Sub-Committee also conducted a field trial of the War Department mess tin ration on men from an air landing brigade. The subjects camped for forty-eight hours and were supposed to eat only the food that they were given. However, after the experiment, researcher Ian A. Anderson reported that 'in a few cases the men did supplement their ration against orders during the exercise by purchases of bread etc. from bakers' roundsmen and by partaking in refreshments offered by civilians.'[99] Resistance could be more outright still. The right of observers to withdraw from trials at Porton was not explicitly mentioned in official regulations until 1962.[100] However, it appears that men could, and did, drop out of experiments during the Second World War. In his report on the preliminary gas-chamber trials with arsenical smokes in 1942 Mackworth noted:

> When the subjects had nearly all finished their second run of the tests, i.e. at zero + 19 minutes, subject 2 who had completed the second series with the classification test began muttering about 'not having bargained for this' and then started to grouse to subject 5 who had just finished the second series at the exercise test. Captain Curwen reassured these men that conditions would soon improve but subject 2 insisted on leaving, followed by subject 5. A few seconds later subject 3 also left when, with the spotter test, he too had finished the second run of all the tests.[101]

During the subsequent assault course field trials of the smokes, researcher Captain H. Cullumbine of the RAMC observed similar behaviours in three men. One 'could not be persuaded to come to the starting point'. Two others attempted the course but quickly dropped out. Cullumbine reported that 'Subject D left the cloud after only 15 seconds. He was lacrymating [shedding tears] but complained of nothing else. He declared that he "was not going to do it". He refused to re-enter the cloud although he was not noticeably affected.'[102] Indeed, resistance could be outright and corporate. Having already done a stint at Porton Down, Wilfred Hall's battalion knew what lay ahead of them. As a result, they refused to do another when the lieutenant ordered them to. Harold recalled that 'The whole section of us was on parade, the whole battalion, and Lieutenant Costello was there inspecting us, and he says at the finish, "Company dismissed," and we all stood fast, never moved. The officers as well, they didn't move either, and he says, "What's the matter?" We says, "Well, we're not going into the gas chambers no more." After this incident we got moved to Bulford. That's where all the naughty boys went.'[103] In these instances the men asserted control over their own bodies by refusing to participate or dropping out of human trials when they felt their health and safety were being compromised, or they had not 'bargained for' what transpired. It seems that they too were guided by a set of underlying assumptions about how their bodies could be used, and when they felt that this had been transgressed, they withdrew from the experiment process.

Moreover, it was not only the soldiers who took part who raised objections to experimentation. The Ministry of Defence reported that during the war there was acrimony between Porton Down and the units providing volunteers about the fitness of men returning from the establishment after testing dressings for mustard gas burns.[104] In 1940 Lieutenant-Colonel F.A. Spencer, commanding officer of the 55th Training Regiment, Royal Armoured Corps, wrote to Army Headquarters at Aldershot complaining that six of his men who had participated in mustard gas experiments returned suffering from blisters on their arms and thighs which varied in size from a half-crown to the size of the palm of the hand. He stated that 'These men have suffered considerable discomfort, and it is considered only just that they should receive some little compensation in addition to the week's leave which has already been granted to them; especially in view of the fact that the subject of compensation was mentioned to them by a sergeant at Porton. I would point out that these men are not complaining in spite of the fact that the blisters in many cases are not yet healed.'[105] Despite the damage to their bodies, it appears that there was no objection by the men, who seem to have accepted their injuries. Rather it was their commanding officer who expressed a sense of outrage. As such, the military body was not only a site of contest between scientist and subject but was located within a more complex set of relations that influenced how it was obtained, experimented on and released back into service.

Conclusion

Between 1939 and 1945 soldiers' bodies were damaged, restored and pushed to the limits of endurance, all in order to further the demands of science and the state. As had been the case in the First World War, the British Army came to liaise closely with the UK's medical and scientific communities, and military officials collaborated with civilian researchers. Through agencies like the Military Personnel Research Committee and the Chemical Defence Experimental Establishment at Porton Down, both military effectiveness and scientific progress were pursued. In order to make possible experiments with, and trials of, new technologies and substances a regular supply of bodies was needed and the army supplied them for most of the war.

It seems that efforts were made to make experiments ethical, but frequently they were not. Nothing beyond an informal exchange of memos ever occurred to set up a system to guarantee that men knew what they faced and that they consented to it. Ethical participation was often thought desirable, but it was never deemed essential. Thus, those designing and running the trials, and those within the military charged with finding volunteers, were unaware of, or chose to ignore, official advice. Within this context, it is difficult to draw conclusions as to why men participated in human trials. Clearly, some were ordered to do so and felt that they had no choice but to obey. Military command structure therefore lent itself to human experimentation as bodies were obtained simply by ordering soldiers to take part.

Yet many soldiers were active participants in the trials on their bodies. Offered the chance to join up voluntarily, they chose to do so for a number of reasons. These included a range of formal incentives such as cash payments, better food and additional leave. These incentives suggest that formal command structures were not enough. The authorities clearly felt that it was necessary to coax troops into experiments rather than simply order them to participate. However, men could also find their own reasons to volunteer, which the authorities may have been unaware of, such as the pleasure of competition, the testing of one's masculinity, taking on new roles and doing one's duty. Perhaps more important still, there are instances where men resisted the demands of their officers and refused to participate, or where they participated but did so in incomplete or unsatisfactory ways. They seem to have faced little sanction. As such, soldiers too had certain expectations for their bodies, and for what the authorities could do to them.

Notes

1 Committee of the Privy Council for Medical Research, *Report of the Medical Research Council for the Year 1938–1939* (Cmd. 6163), p. 14.
2 S. Sturdy, 'War as experiment: physiology, innovation and administration in Britain, 1914–1918', in Cooter, Harrison and Sturdy (eds.) *War, Medicine and Modernity*, pp. 74–5.
3 *Medical Research in War: Report of the Medical Research Council for the Years 1939–1945* (Cmd. 7335), p. 13.
4 See, for example, J. Moreno, *Undue Risk: Secret State Experiments on Humans* (London: Routledge, 2000); S. Lederer, 'Military personnel as research subjects', in S.G. Post (ed.), *Encyclopedia of Bioethics* (London: Macmillan, 3rd edn, 2004), p. 1843; M.E. Frisna, 'Medical ethics in

Experimentation

military biomedical research', in T.E. Beam and L.R. Sparacino (eds.), *Military Medical Ethics, Volume II* (Falls Church, VA: Office of the Surgeon General, Department of the Army, 2003), pp. 533–61; L. Black, 'Health law: informed consent in the military: the anthrax vaccination', *American Medical Association Journal of Ethics* 9:10 (2007), 698–702; J. McManus et al., 'Informed consent and ethical issues in military medical research', *Academic Emergency Medicine* 12:11 (2005), 1120–6.

5 *Medical Research in War* (Cmd. 7335), p. 135.
6 *Medical Research in War* (Cmd. 7335), p. 135.
7 H.E. Holling, 'Wartime investigations into sea and air-sickness', *British Medical Bulletin* 5:1 (1947), 46–7.
8 *Medical Research in War* (Cmd. 7335), p.142; W.S.S. Ladell, 'Effects on man of high temperatures', *British Medical Bulletin* 3:1 (1947), 5–6.
9 Ladell, 'Effects on man of high temperatures', 6.
10 Ladell, 'Effects on man of high temperatures', 8.
11 Lt.-Col. W. Somerville, 'The effect of Benzedrine on mental or physical fatigue on soldiers', *Canadian Medical Association Journal* 55 (1946), 55.
12 TNA FD1/7064, Fifth meeting of Sub-Committee on Analeptic Substances (Benzedrine), 28 February 1942, appendix 1.
13 *Medical Research in War* (Cmd. 7335), p. 140.
14 D.P. Cuthbertson and J.A.C. Knox, 'The effects of analeptics on the fatigued subject', *Journal of Physiology* 106 (1947), 42–58.
15 TNA FD1/7064, Seventh meeting of Sub-Committee on Analeptic Substances, 17 July 1942, p.1; Cuthbertson and Knox, 'The effects of Benzedrine on the fatigued subject', 54.
16 TNA WO203/691, Medical report on the use of Benzedrine by armoured troops, December 1942.
17 TNA WO222/97, Notes on the use of Benzedrine in war operations, 23 December 1942, p. 1.
18 *Medical Research in War* (Cmd. 7335), p. 21.
19 J. Brozek, 'Psychological war-time research in Great Britain', *American Journal of Psychology* 62 (1949), 125.
20 *Medical Research in War* (Cmd. 7335), pp. 145–6.
21 L.A. Legros and H.C. Weston, 'On the design of machinery in relation to the operator', *Industrial Fatigue Research Board, Report No. 36* (London: HMSO, 1926).
22 R.S.F. Schilling, 'Industrial health research: the work of the Industrial Health Research Board, 1918–1944', *British Journal of Industrial Medicine* 1 (1944), 148–9.
23 Schilling, 'Industrial health research', 132.
24 Schilling, 'Industrial health research', 138.
25 *Medical Research in War* (Cmd. 7335), p. 136.
26 Ministry of Defence (hereafter MOD), Historical Survey of the Porton

Down Service Volunteer Programme, 1939–1989 (2001), http://webarchive.nationalarchives.gov.uk/20060715135118/http://www.mod.uk/DefenceInternet/AboutDefence/Issues/HistoricalSurveyOfThePortonDownServceVounteerProgramme19391989.htm, pp. 32–4 (accessed 21 November 2013).

27 TNA WO286/11, Volunteers for physiological tests at Chemical Defence Experimental Establishment, Porton, 5 June 1953, Minute sheet 1.
28 TNA WO286/11, Volunteers for physiological tests at Chemical Defence Experimental Establishment, Porton, 5 June 1953, Minute sheet 1.
29 Moreno, *Undue Risk*, p. 37.
30 TNA WO189/2293, N.H. Mackworth, A psychological test for the harassing effects of lachrymators on vision, 3 June 1942; H. Cullumbine, 'Chemical warfare experiments involving human subjects', *British Medical Journal* (19 October 1946), 576–8.
31 TNA WO189/2270, Captain S. Curwen, Medical report on casualties produced by airburst mustard gas shell, 10 March 1942, p. 1.
32 MOD, Historical Survey, p. 245.
33 MOD, Historical Survey, p. 285.
34 IWM SA, 17925, Stanley Shore, reel 1.
35 TNA WO189/2270, Curwen, Medical report on casualties produced by airburst mustard gas shell, p. 2.
36 D.C. Sinclair, 'Disability produced by exposure of skin to mustard gas vapour', *British Medical Journal* (11 February 1950), 346–8.
37 MOD, Historical Survey, p. 23.
38 MOD, Historical Survey, p. 23.
39 TNA WO189/2293, Mackworth, A psychological test for the harassing effects of lachrymators on vision, p. 2.
40 TNA WO189/2293, Mackworth, A psychological test for the harassing effects of lachrymators on vision, p. 4.
41 W. E. Le Gros Clark, 'The contribution of anatomy to the war', *British Medical Journal* (12 January 1946), 39.
42 *Medical Research in War* (Cmd. 7335), p. 28.
43 *Medical Research in War* (Cmd. 7335), pp. 188–9.
44 *Medical Research in War* (Cmd. 7335), pp. 33–4.
45 *Medical Research in War* (Cmd. 7335), p. 54.
46 TNA WO222/1543, Traumatic Shock Team, Report for the quarter ending 31 March 1945, including a review of the year's work in the field, p. 3.
47 P. Neushel, 'Fighting research: army participation in the clinical testing and mass production of penicillin during the Second World War', in Cooter, Harrison and Sturdy, *War, Medicine and Modernity*, p. 204.
48 Neushel, 'Fighting research', pp. 85–6.
49 'Supplies and distribution of penicillin: statement by the Medical Research Council', *British Medical Journal* (23 August 1943), 274.

Experimentation

50 *Medical Research in War* (Cmd. 7335), p. 58.
51 'Supplies and distribution of penicillin: statement by the Medical Research Council', *British Medical Journal* (23 August 1943), 274.
52 IWM SA, 19674, Basil Reeve, reel 3.
53 'Supplies and distribution of penicillin: statement by the Medical Research Council', *British Medical Journal* (23 August 1943), 274.
54 'Supplies and distribution of penicillin: statement by the Medical Research Council', *British Medical Journal* (23 August 1943), 274.
55 *Medical Research in War* (Cmd. 7335), p. 87.
56 IWM SA, 19674, Basil Reeve, reel 3.
57 *Medical Research in War* (Cmd. 7335), p. 79.
58 J.W.S. Pringle, 'Effects of World War II on the development of knowledge in the biological sciences', *Proclamation of the Royal Society of London*, A. 342 (1975), 539.
59 TNA WO286/11, Volunteer observers for physiological tests at Porton, 29 July 1944, Minute sheet 1.
60 TNA WO286/11, Volunteer observers for physiological tests at Porton, 29 July 1944, Minute sheet 1.
61 TNA WO286/11, Volunteer observers for physiological tests at Porton, 29 July 1944, Minute sheet 1.
62 K. Mellanby, *Human Guinea Pigs* (London: Merlin, 1973), pp. 58–9.
63 *Medical Research in War* (Cmd. 7335), pp. 58–9.
64 *Medical Research in War* (Cmd. 7335), p. 67.
65 The concept of the moral economy was developed by E.P. Thompson in 'The moral economy of the English crowd in the eighteenth century', *Past and Present*, 50:1 (1971), 76–136. See also E.P. Thompson, *The Making of the English Working Class* (London: Gollancz, 1963), pp. 62–3.
66 J. Goodman, A. McElligott and L. Marks, 'Making human bodies useful: historicizing medical experiments in the twentieth century', in J. Goodman, A. McElligott and L. Marks (eds.), *Useful Bodies: Humans in the Service of Medical Science in the Twentieth Century* (Baltimore: Johns Hopkins University Press, 2003), p. 3.
67 TNA TS27/398, Treasury Solicitor to Medical Research Council on the question of the legal position of the Council if volunteer subjects for proposed experiments should die, 21 June 1933.
68 TNA WO189/2848, Wing Commander E,C.B. Bramwell, History of the Service Volunteer Observer Scheme at Chemical Defence Experimental Establishment (CDEE), 27 November 1959, pp. 2–3.
69 TNA WO286/11, Volunteer observers for physiological tests at Porton, Minute sheet 1; MOD, Historical Survey, p. 23.
70 TNA WO286/11, Volunteer observers for physiological tests at Porton, Minute sheet 1.

71 TNA WO289/11, Letter from War Office to Air Ministry, 28 October 1932; Letter from War Office to Admiralty, 28 October 1932.
72 TNA WO189/2848, Memorandum by the War Committee, Employment of observers from the services on physiological tests involving exposure to sternutators, 23 April 1940, p. 12.
73 TNA WO189/2848, Wing Commander E.C.B. Bramwell, History of the Service Volunteer Programme at CDEE, 27 November 1959, p. 16.
74 MOD, Historical Survey, p. 23.
75 R. Evans, *Gassed: British Chemical Warfare Experiments on Humans at Porton Down* (London: House of Stratus, 2000), p. 6.
76 The decision to include only those men who had been part of the police investigation was based on the fact that they had made their details available. While the identities of all servicemen who attended Porton were recorded in the experimental logs, access was restricted under data protection. Of the men who were contacted, 401 replied. They were not responding to a particular grievance but to a general enquiry about their visits. MOD, Historical Survey, p. 469.
77 MOD, Historical Survey, p. 483.
78 MOD, Historical Survey, p. 483.
79 IWM SA, 16084, Edward Kirby, reel 5.
80 IWM SA, 22072, Wilfred Hall, reel 1.
81 No exact statistics are available for the Second World War.
82 MOD, Historical Survey, p. 314.
83 MOD, Historical Survey, p. 483.
84 Evans, *Gassed*, p. 5.
85 Holling, 'Wartime investigations into sea and air-sickness', p. 47.
86 TNA WO189/2321, N.H. Mackworth, The use of an assault course in the assessment of the arsenical smokes, 18 August 1942, p. 15.
87 TNA WO189/2321, Mackworth, The use of an assault course in the assessment of the arsenical smokes, p. 3.
88 Holling, 'Wartime investigations into sea and air-sickness', p. 47.
89 Mellanby, *Human Guinea Pigs*, pp. 79–80.
90 TNA WO189/2848, History of the Service Volunteer Observer Scheme at CDEE, 27 November 1959.
91 IWM SA, 17925, Stanley Shore, reel 1.
92 IWM SA, 30407, Ralph Kirkton, reel 4.
93 MOD, Historical Survey, p. 30.
94 *Pay and Allowances of the Armed Forces, 1941–42* (Cmd. 6385), p. 2.
95 IWM SA, 17925, Stanley Shore, reel 1.
96 IWM SA, 16084, Edward Kirby, reel 5.
97 BBC WW2 People's War Archive, A2669222, Harold Pollins, 26 May 2004, www.bbc.co.uk/history/ww2peopleswar/stories/22/a2669222.shtml (accessed November 2013).

Experimentation

98 Holling, 'Wartime investigations into sea and air sickness', 48.
99 TNA FD1/7042, B.P.C. 42/16/FR27, Report on a field trial of War Department mess tin ration, 19 March 1942, p. 1.
100 MOD, Historical Survey, p. 314.
101 TNA WO189/2321, Mackworth, The use of an assault course in the assessment of the arsenical smokes, appendix 1: Details of preliminary chamber trials, 1942, p. 15.
102 TNA WO189/2321, Captain H. Cullumbine, Comments on the behaviour of the subjects during and immediately after exposure to the cloud from D.A. Generators, May 1942, pp. 24–5.
103 IWM SA, 22072, Wilfred Hall, reel 1.
104 MOD, Historical Survey, p. 24.
105 TNA WO188/1449, Letter from Lieutenant-Colonel F.A. Spencer, Commanding 55th Training Regt. Royal Armoured Corps, 29 November 1940.

4

Active service

By 1945 over two million British Army personnel were serving overseas, including in parts of South Asia, North Africa, Europe and the Mediterranean.[1] The civilian-turned-soldier, whose body had been rendered fit, ordered and productive through training, was now to fulfil his military obligations in the field of active operations. Fighting the enemy was not, however, his only challenge. The army *Handbook of Military Hygiene*, published by the War Office in 1943, stated that 'the best of equipment and training will be of little avail if men fall sick before they can put their training and skill at arms into practice.'[2] In fact, in all theatres of war, sickness was a greater problem in terms of manpower than battle casualties.[3] A key task for the military authorities was therefore to maintain the health and efficiency of troops in various combat zones. This continued military management of the body within the field of active service will be explored in this chapter.

The army's medical arrangements in the field have been examined in detail elsewhere.[4] It is not the intention of this chapter to replicate this work, but to put the body at the heart of the analysis. In doing so, it considers links between the military body of the Second World War and British imperial tradition. In his work on the Western Desert, Mark Harrison has shown that the British Army's experience of fighting in hot climates gave them a certain edge over their enemies. While Germany had acquired few colonies, the British had regularly despatched expeditionary forces to fight in hot climates and had maintained large overseas garrisons, most notably in India. As a result of bitter experience there, they had evolved strict rules and regulations governing matters of hygiene and sanitation.[5] These rules and regulations, I suggest, reflected a conception of the human body as environmentally sensitive, an idea that had long been part of Western medical thinking and had led to a range of efforts to physiologically adjust Europeans to their new surroundings. The army's attempts to safeguard the health of British men also reflected a conception of the British body as distinct from the 'other' bodies that it encountered, particularly those of indigenous populations, which were perceived as infectious sites of disease.

What is also explored in this chapter is the nature of the control upon men's bodies on active service. John Hockey suggests that outside the training environment, in operational units, bodily regulation becomes relatively relaxed. Yet, as a result of the internalisation of discipline during training, the operational body, 'when exhausted, freezing and fearful, will still constantly monitor itself (and those of its peers)'.[67] This chapter shows that the British Army of the Second World War did not simply rely on training back at home to instil self-discipline in its men. Rather, various additional strategies were designed specifically for the field of operations. These encouraged men to continue to look after their own bodies even in the harsh conditions of warfare. These methods, and soldiers' reactions to them, will be examined in order to consider the 'released' body's response to the loosening of formal organisational control.

Bodies and environments

Men deployed to active service overseas faced a range of new assaults upon their bodies. Even before the threat of enemy action, they encountered a variety of new diseases and health issues. Troops sent to tropical climates, for example, were liable to suffer from malaria, heat stroke and tropical neurasthenia, while those deployed into arctic conditions were threatened with frostbite, trench foot and snow blindness.[8] Highlighting the vulnerability of the British body to the changed environment, Lieutenant-Colonel E.T. Burje advised soldiers in a 1942 publication, *Tropical Tips for Troops*:

> Your human body, born and brought up in a temperate, not to say chilly, climate, is really a very delicately adjusted piece of mechanism. It has for some twenty or more years become habituated to the meteorological and nutritive conduits obtaining in the United Kingdom...Any sudden change from one set of conditions to an entirely different one, throws an immense strain upon the physical and mental mechanism.[9]

This attitude towards the body and its relationship with the environment was not new to the British Army of the Second World War, but echoed earlier colonial anxieties. Territorial expansion, particularly in South

Asia, meant that the British had long been concerned about the impact of changed conditions on European bodily constitutions. Military and medical experts had thus formulated a variety of rules and regulations in order to survive in hostile environments.[10] Between 1939 and 1945 the military authorities drew upon this experience to implement a range of strategies to maintain the health and efficiency of overseas troops. In doing so, they came to conceptualise the human body itself in two distinct, though not mutually exclusive, ways.

Although it was the product of the home environment, and more naturally suited to it, the British body was, nevertheless, conceived as an open, fluid entity that to was able to adapt to new surroundings. The *Handbook of Military Hygiene* advised that 'the human body is in many ways similar to the engine of a motor-car. It requires fuel and water, its temperature is regulated by a cooling system, and its waste products must be removed. It differs, however, in the fact that it carries out its own repairs, and that it readily responds to changes in its environment.'[11] In fact, physiological make-up was considered as fundamentally the same across geographical boundaries, differing only in terms of adaptability. An article in the *Journal of the Royal Army Medical Corps* noted in 1944:

> As far as climatic optima – temperature, humidity, atmospheric movement, sunlight – are concerned, as judged by their effects upon human efficiency in performance, there is no difference of any great magnitude between the different geographical varieties of mankind. It is possible for all these varieties to flourish biologically in all parts of the world. Nevertheless the time required for adjustment to a new and different set of climatic conditions varies with different individuals and types, ages and previous experiences. Thus, though we as a people are more in tune with the physiological and meteorological conditions of our own country than we are with environments markedly different from this, it is still eminently possible for us to venture into any part of the world where the war may take us and therein adapt ourselves to its climatic conditions.[12]

As something that was permeated and changed by the environment, the body could therefore be adjusted to it. To this end, the army adopted several techniques. A popular method used to prepare men for service in tropical and desert climates was the use of 'controlled sunbathing' aboard ships. A report by the Army Directorate of Medical Services explained that this allowed for 'some degree of skin tanning which is of great assistance in promoting healthy skin, and the reduction of prickly heat'.[13] Lieutenant-Colonel Burje even encouraged men to 'indulge in sun-bathing parades'. He recommended that troops should wear progressively fewer clothes over three days before 'wearing a helmet only, the whole body should be exposed to the sun. It is best to go completely nudist. If you or your men are too modest for that – a thing you will soon get over – then wear only whatever is the modern equivalent for a fig leaf.'[14] Indeed, communal sunbathing often became a highly social event, where men played card games with each other and participated in activities like tug-of-war.[15] Harry Blood, who travelled aboard ship from Glasgow to Egypt in April 1941, wrote in his diary, 'lying here in the sun with the crowd sunbathing and playing 'housie-housie' reminds me of summer holidays.'[16] Upon arrival at their destinations, soldiers then underwent a three- to six-week acclimatisation period in base areas.[17] Known during the Second World War as 'salting', this technique had been practised by colonial settlers since the late eighteenth century.[18] Men deployed to the Middle East, for example, spent their first six weeks in the Canal area of Egypt. Colonel A.E. Richmond and Lieutenant-Colonel H.S. Gear, deputy and assistant directors of hygiene, described the benefits of the 'salting' process:

> This allowed physiological and, frequently, though not by design, an immunological adaptation. On the first count, the body and mind became attuned to heat, glare, dust and the harsh environment of vast desert landscapes. Secondly, in spite of care, many newly arrived units suffered from enteritis, sandfly fever and sunburn. Occurring in the settled conditions of base camps, not much harm resulted. If, however, troops had suffered these disabilities in action, serious consequences to the strategy and tactics of the forces involved might have followed. A 'salting' process occurring in base camps was, therefore, not altogether a disadvantage.[19]

This early period of exposure thus allowed the body to be safely and gradually adjusted to the hot climate and to develop natural immunity against new disease pathogens. As Richmond and Gear explained, this was crucial to maintain the performance of the troops when deployed into active operations.

A more artificial method of manipulating the body's natural defences was the use of 'preventative inoculations', which, by introducing dead germs into the body, caused the formation of protective substances within the blood.[20] The *Journal of the Royal Army Medical Corps* described inoculations as 'amazing feats of biological magic by which the human body is educated and trained to defend itself'.[21] All men were inoculated against tetanus and the typhoid groups of fevers before going abroad and against cholera or plague whenever an outbreak was threatened. Inoculation against yellow fever was also made available for troops in countries where the disease was a threat.²² The efficacy of vaccination was demonstrated among prisoners of war (POWs) in the Western Desert during outbreaks of typhoid between 1941 and 1942. While Italian prisoners in a British POW camp had received the Italian TAB (typhoid-paratyphoid A and B) vaccine and experienced numerous cases of the disease, British prisoners in an enemy camp, who had been were inoculated with the British vaccine, suffered no cases of typhoid. This was despite poor sanitary conditions, outbreaks of dysentery and cases of typhoid disease among enemy troops in the neighbourhood.²³

In order to adapt men's bodies nutritionally to different environments, the army also developed a wide range of field ration scales. The *Handbook of Military Hygiene* explained that 'the quantities are regulated by the soldier's needs which may vary according to circumstances...The greater the amount of muscular work done, the more food required. In cold climates more food is needed to keep the body warm.'²⁴ The nutritional and calorific values of the various ration packs were impressively precise. Men serving in the Central Mediterranean Force, for instance, received 3,600 calories per day, while those in South-East Asia were provided with 4,500 calories.²⁵ The Pacific 24 Hour Ration (British Troops), Pacific Compo Ration and Pacific Emergency Ration were designed specifically for jungle and tropical warfare and provided a highly nutritive and varied ration that took up the minimum of space and weight. The Mountain (Arctic) Pack Ration was created specifically for cold and isolated environments. It contained dehydrated and tinned foods and had a calorific value of 5,100. When fresh supplies were hard to acquire, men were given vitamins in the form of marmite, yeast tablets and ascorbic acid tablets. From 1943 troops also received a compound vitamin tablet. This strict attention to dietary requirements meant that by the end of the war there were forty-nine separate ration scales designed to meet the nutritional needs of men in different theatres of war.²⁶

While the army clearly devoted much effort to adjusting bodies to new environments, it also took action to distance these bodies from those environments. This reflected an alternative conception of the body as somewhat biologically fixed and limited in its ability to adapt. The *Journal of the Royal Army Medical Corps* noted that among the most valuable contributions of the Army Medical Services was 'the reinforcement, by artificial means, of the defensive mechanisms of these individuals, so that they might thereby be more fully protected against the disease-provoking agencies which they are likely to encounter in the environments to which they were proceeding'.²⁷ These 'defensive mechanisms' included

Active service

protective clothing, personal prophylaxis, insect repellents and mosquito nets. Highlighting the importance of the latter, Lieutenant-Colonel Burje advised men that before leaving Britain they should 'purchase some two yards of netting and get some obliging girl friend to make it into a bell-shaped structure which can be placed over the top of the helmet so as to hang down all round.'[28] Soldiers serving in hot climates were also issued with light-coloured cotton, linen or silk clothing. These materials had fewer air spaces, were better conductors of heat and were considered to be cooler and more comfortable.[29] In parts of the Middle East the wearing of slacks instead of shorts was even made obligatory in order to protect men against insects, desert sores and burns.[30] Troops serving in cold climates, on the other hand, were provided with woollen clothing, which had the largest proportion of air spaces, absorbed water quickly and dried slowly to prevent the chilling of the body by too rapid evacuation when soaked with rain or sweat.[31]

Safeguarding the soldier's body was also a spatial process; whenever possible, the authorities kept men away from areas deemed suspect. In the Middle East, military camps were located away from local villages and towns, which were identified as the main sources of fly-breeding. Malaria field laboratories surveyed the whole area and classified areas into 'highly malarious', 'malarious' and 'non-malarious'. Military installations were then limited to the non-malarious regions.[32] Evaluating landscapes in terms of their effects upon health, in fact, became a significant part of military intelligence during the Second World War through the work of the Inter-Service Topographical Department. This special unit consisted of over 700 British and Allied personnel, who provided detailed reports and maps that assessed overseas terrain to guide the planning of military operations.[33] The intelligence gathered included information on the diseases that were most prevalent in certain areas and the hygienic arrangements that needed to be put in place. In a 'Climate and Medical' report on southern Italy in June 1943, for example, investigators noted that below the Gaeta–Foggia line (which included the provinces of Campania, Lucania, Calabria and Puglia), malaria and blackwater fever were extremely common. Calabria was reported as the most endemic area with 185 deaths per season (May–December). Based on this information, the investigators recommended the avoidance of such areas where possible and personal prophylaxis when it was not. Sandfly fever was also reported to be abundant from May to October so the department advised that 'camp siting should be up-wind from breeding places.'[34] In light of these endeavours, it seems that bodies themselves became barometers of place, as the authorities sought to manage geographic locations through the diseases that they produced.

It was not, however, always possible or practical to separate bodies from hazardous environments, so the army also made efforts to protect British men by conquering the landscapes that they encountered. The *Journal of the Royal Army Medical Corps* noted the need for 'the banishing of disease provoking agencies from these environments or the achievement of a maximum degree of control over these agencies.'[35] Most of this work was carried out by the Army Hygiene Service, which was responsible for 'the supervision in general of the environment of the soldier from the point of view of the preservation and enhancement of his health and fitness'.[36] On the ground, field hygiene units and sanitary sections cleansed overseas theatres of disease-spreading dirt, human waste and insects. In the Western Desert, for instance, special units burned refuse, treated contaminated waters and removed human waste. A special Fly Control Unit was also created to clear the whole El Alamein defence line of fly-breeding material, such as dead bodies, litter and refuse.[37] The Hygiene Service also worked to overcome problems with water supplies so that soldiers could keep themselves clean and free from infection. In Egypt purified water was transported to front-line troops through a comprehensive system that extended along the camps close to the Suez Canal and into the desert area to the north of the Sweet Water Canal.[38] During an assault on the Assam border behind enemy lines in Burma, clean water was also flown in for Allied troops.[39] Clearly, the military authorities of the Second World War did not simply try to protect the body by focusing on that body. Rather, the environment itself was regulated in order to mitigate the threat posed.

British bodies and dangerous 'others'

If the local environments that British soldiers lived and fought in were thought to represent a menace to their bodies, so too were the dangerous 'other' bodies that the men encountered. Official concerns focused particularly on enemy troops and indigenous populations, whose poor sanitary habits were believed to be a key cause of disease. In a report in the *British Medical Journal* on the sanitary conditions at the Battle of El Alamein, Gear noted that 'the countryside had been thickly populated with Germans, Italians and Arabs, and was now rank with pollution. It is understandable that a fast-retreating and defeated army cannot bury its dead. But it is always surprising to find that German soldiery fouls its own nest with a thoroughness that is truly Teutonic.'[40] This opinion was echoed by a hygiene officer at the battle who was quoted as saying, 'that portion of the battlefield previously occupied by the enemy is just one huge fly farm and has to be seen to be believed. Whilst both Germans and Italians order the use of shallow trench latrines (and no oil seal), this order is scarcely ever carried out. Enemy defensive localities are obvious from the amount of faeces lying on the surface of the ground.'[41] These accounts suggest that the environment that was so harmful to the British soldier was itself man-made, the result of occupation by suspect other bodies. In this instance, German and Italian troops were the offenders. Their poor sanitary instincts were blamed for generating the filth that would give rise to disease. Military officials even believed that the contamination of the landscape was a deliberate tactic used by the enemy against Allied troops. Reports by medical topographers for southern Italy noted that under the Fascist regime there had been 'marked improvement in all sanitary matters, especially the control of malaria', including the establishment of malaria committees and the drainage of large areas of marshland. The investigators advised, however, that much of this work had been abandoned during the war and it should be expected that 'measures would be taken during a retreat to leave an area, previously healthy in terms of malaria, as much as possible a danger to an invading force'.[42]

Unhealthy sanitary habits were thought to be even more inherent within the native communities that British troops encountered. Describing the army's retreat from Gazala in 1942, Gear noted that, 'Apart from the disturbance the retreat caused in the hygiene organization, it created a new problem by sweeping back hordes of natives, Bedouins, and others, to settle in an uncontrolled mass in the rear of the El Alamein line. These natives, normally lacking any high sanitary instincts, became a serious menace to the Army, and drastic action had to be taken to have them removed to harmless areas further back still.'[43] Moreover, the endemic nature of certain diseases within indigenous populations meant that the native body itself was conceived as a particularly infectious site. The *Health Memoranda for Troops in the Tropics* stated that 'unfortunately in the tropics, we live amongst people many of whom are "carriers" of certain common and important diseases. These carriers are capable of spreading their diseases to the healthy man near them.' It continued, 'several diseases and conditions, which are uncommon at home, are very common amongst the population, wherever we go. In our daily life we are surrounded by native personnel, many of whom are carriers of those diseases.'[44] As such, the indigenous body was portrayed as a silent threat, secretly harbouring the diseases that threatened British troops.

Once again, the army took a two-pronged approach when dealing with the health risks posed by native communities. Echoing older colonial practices of siting military cantonments away from civilian populations, troops were typically kept away from civilian populations.[45] The stationing of army camps outside towns and villages distanced men's bodies not only from the threat of insects and dirt, but also from the human sources of disease. Richmond and Gear explained that it reduced contact with 'such infections as typhus, smallpox, plague and venereal disease. These diseases in the military in the Middle East had a close correlation with civil urban communities.'[46] The British also, however, focused on reforming the indigenous communities that they encountered. These strategies in part reflected older traditions of colonialism and the 'civilising mission' of empire.[47] In the Western Desert all native labourers and POWs were routinely disinfected.[48] In the occupied enemy territories of Eritrea, Cyrenaica and Tripolitania in the Middle East the British carried out medical and health supervision and improved the sanitary arrangements in urban communities. Medical officers and health inspectors also implemented water services, hospitals, dispensaries and clinics. Richmond and Gear noted that as a result of these measures the medical services had, 'within a year, got the gross filth of generations removed'.[49]

There was one final specific threat to British men serving overseas: the native female body, which was identified by the authorities as a potent source of venereal infection. This was a serious drain on manpower in many theatres abroad. In the Middle East Force an average of 28 men in every 1,000 contracted venereal disease during the war. In India, the figure stood at 66. Before the introduction of penicillin, a soldier who was infected with gonorrhoea usually spent 25 days in hospital, while a man with syphilis would be out of action for between 40 and 50 days.[50] The problem, as medical officials saw it, was that the army often had to contend with an extensive legalised brothel system as well as numerous amateur or 'clandestine' prostitutes.[51] These women were held responsible for the spread of disease. The *Health Memoranda for British Soldiers in the Tropics* advised that 'promiscuous sexual intercourse is bound to bring certain risk of infection with syphilis or gonorrhoea (clap). It can be taken for granted that any native woman who solicits your attention is, or has been infected with one or another, or both, of these diseases.'[52] In Italy, where venereal disease caused more casualties than battle wounds after the Allied occupation in 1943, Director of Medical Services Major-General E.M. Cowell proclaimed that that 'Italian women are causing as much damage to the Army as the German men. Venus has become a Delilah (V.D.).'[53] He reported that because of the economic situation women were offering themselves for a few cigarettes and that 'prostitutes abound'. Nearly all of these women were suffering from a virulent form of the disease, meaning that the risk of infection to men 'who have one of these women' was 80 per cent.[54]

This demonising of the female body again had earlier origins and reflected traditional military attitudes and gender stereotypes. In the nineteenth century, British military leaders both at home and in the colonies had been convinced that soldiers could not possibly curb their own sexual desires or that any attempts to do so might undermine their heterosexuality, which the army prized so highly. As such, the spread of venereal infection had been blamed solely on the women who solicited men's attention.[55] Anxieties about venereal disease had been particularly manifested in debates about the Contagious Diseases Acts. By allowing for the identification, treatment and detention of prostitutes within major garrison towns, this legislation had isolated women as the source of infection.[56] According to Lucy Bland, this notion of 'woman as polluter' continued to be commonplace into the twentieth century as official reports and memoranda repeatedly stressed the need to target prostitutes and 'promiscuous amateurs' who consorted with servicemen.[57] As the number of service cases in Britain rose during the Second World War, a new system of compulsion was brought in under Defence Regulation 33B.[58] Persons named by two patients under treatment for venereal disease as the source of their infection could be compelled to attend for examination and, if necessary, treatment by a 'special practitioner'.[59] While this law was not targeted specifically at women, Roger Davidson has argued that the rationale behind it was still highly discriminatory. Although the government claimed that the new controls were designed to protect 'all those engaged in essential war work', official draftsmen clearly assumed that the bulk of informants would be servicemen, and that sexually precocious girls and promiscuous women would usually be notified as the source of infection.[60]

It is perhaps not surprising, then, that British military officials overseas sought to safeguard the soldier's body by regulating female sexual behaviour. In an attempt to reduce numbers of clandestine prostitutes in France, for example, the director of medical services of the British Expeditionary Force adopted a policy of formally identifying women who were known to be spreading venereal disease amongst troops.[61] In order to control clandestine prostitution and 'pimping' in the Mediterranean, a team of British senior medical officers, led by consultant venereologist Brigadier Robert Lees, recommended that, 'if necessary, laws must be framed imposing extremely heavy penalties, including imprisonment, for such offences, and such penalties to be widely reported in the local press.'[62] In parts of France, the Mediterranean and the Middle East, the army also established systems of regulated brothels where medical officers or municipal doctors inspected prostitutes for venereal disease.[63] In keeping with the traditions of military rank and hierarchy, separate establishments were created for the rank and file and for those with the King's Commission. Describing a brothel for ordinary troops in Italy, infantryman James McCallum wrote in his memoir:

> The army, with its detailed administrative ability, was able to organise brothels in a surprisingly short time and a pavement in Tripoli held a long queue of men, four deep, standing in orderly patience to pay their money and break the monotony of desert celibacy. The queue was four deep because there were only four women in the brothel. The soldiers stood like units in a conveyor belt waiting for servicing. I hope these four persevering women are awarded decorations by the British government.[64]

Men visiting army-run brothels signed their names and army numbers and were given a slip. Any individual found later to be suffering from venereal infection but who was in possession of a slip would not be reprimanded. George Percy visited a regulated brothel Egypt in 1943. He explained that 'if you later got VD and you'd been to the army brothel, you were tret lightly if you had your name down and that, but people that went into the brothel very rarely got VD because it was run by the army and the lasses were inspected every day.'[65] By establishing regulated brothels, the army therefore did not try to curtail men's sexual desires, but focused on providing a safe space in which they could fulfil them. Certainly, it seems that George trusted that the authorities would protect his body by monitoring the women that he encountered.

Policing the female body was, however, only one part of the army's campaign against venereal disease. Officials also attempted to cut rates of infection by regulating the sexual habits of soldiers. In order to encourage men to engage in 'healthy' recreation, the army provided improved facilities like organised sports and leisure activities in some base areas. In several overseas theatres brothels were also placed 'out of bounds' to British troops, and prostitutes were excluded from military zones.[66]

Perhaps the most famous example of this was the Cairo Berka, which was closed to service personnel in 1942.[67] This approach was again not completely new. As Olive Anderson has shown, purity movements and moral reformers had targeted British military personnel since the mid-nineteenth century. A new concept of 'Christian militarism' had emerged that stressed the virtues of chivalry and self-restraint.[68] Such attitudes had continued to play an important part in efforts to prevent venereal disease among troops overseas during the First World War.[69] According to Mark Harrison, attempts to regulate male sexual behaviour during the Second World War now reflected a new conception of citizenship that distinguished the British soldier from his Axis counterparts. The 'typical British male', he suggests, 'was portrayed as free and self-directed, but his freedom (including his sexual freedom) was bound by a sense of responsibility, moderation and 'good form'.'[70] Official discussions about venereal disease certainly drew heavily on the rhetoric of nationhood, patriotism and comradeship. Among the foremost proponents of abstinence were senior medical staff like Major-General Cowell, who commented that 'continence is a duty to oneself, to one's family and one's comrades'.[71] Brigadier Lees similarly remarked, 'I consider control of V.D. a matter of discipline and "morale" much more that of medical measures. Self-discipline comes first, with pride in being fit to fight and fit to serve'.[72] In order to curb the spread of venereal infection in the central Mediterranean, Lees recommended that 'it should be clearly understood by every man that it is a disgraceful act to endanger his health while on active service, by consorting with any loose women. A high code of personal morality must be followed and all must be taught that complete abstinence from sexual intercourse is not detrimental to health or vigour. Association with public prostitutes is "conduct unbecoming an Officer and a Gentleman"'.[73]

To high-ranking officers like Lees and Cowell contraction of venereal disease therefore represented a failure on the part of the individual. Sexual promiscuity was a sign of weakness, neglect and a lack of self-restraint. As Lees' statement shows, this sort of behaviour was considered to be especially dishonourable among the officer class; they were men who were expected to uphold the virtues of chivalry and self-control. He asserted that association with prostitutes was 'conduct unbecoming an Officer and a Gentleman' – the most classic 'honour crime' within the British Army.[74] Although this was not actually defined as an offence in military law, it fell under the category of 'disgraceful conduct' and carried the risk of dismissal. This deliberately broad, all-encompassing category allowed the authorities to discipline officers for behaviours that they considered to be particularly offensive, including fraud, embezzlement, stealing, malingering, self-inflicting wounds and any other 'disgraceful conduct of a cruel, indecent or unnatural kind'.[75] As such, discussions about sexual behaviour appear to have reflected wider general assumptions about class differences, as officers were specifically targeted for moral instruction under military law.

The authorities also, however, took a pragmatic approach when it came to regulating the sexual behaviour of the troops. Recognising that not all soldiers would abstain, the medical services sought to limit the negative consequences of sex by providing treatments and prophylaxis. Lees reported in 1942 that 'man is frail, and the sexual impulses powerful, and *in a minority* risks will be taken and exposure to V.D. occur.' He recommended 'protection during the act' and 'disinfection after intercourse'.[76] A 1943 editorial in the *Journal of the Royal Army Medical Corps* on 'The prevention of venereal disease in the army' also noted that, 'having taught the soldier the necessity for continence, the dangers of promiscuity, what venereal disease is and what it may lead to, there remain the more positive measures which can be employed in those cases where, none the less, men will indulge. In general the policy is to encourage early personal disinfection after possible risk.'[77] In order to reduce the chances of infection, the army issued condoms, commonly known as 'French letters', to troops serving overseas.[78] Men were also supplied with personal prophylactic packets containing some cotton wool impregnated with soft soap, a tube of antiseptic cream and instructions explaining how to use them. Half the antiseptic was to be applied before intercourse, followed by washing with soap and water and the application of the remainder of the ointment after coitus. Such a wide-scale distribution of personal prophylactic kits was unprecedented among the British armed forces during war. While several other armies, such as those of the USA and New Zealand, had given their troops chemical prophylactics during the First World War, the strength of feeling against such a measure had been so strong in the UK that it was not until later in 1918 that British servicemen were issued with prophylactic tubes.[79] An additional measure between 1939 and 1945 that had been implemented during the First World War was the establishment of prophylactic centres in overseas theatres.[80] Here soldiers could either disinfect themselves or be given a 'wash out' by a medical professional. The urethra would be irrigated with an antiseptic such as potassium permanganate before calomel lotion was applied externally. These proved to be popular with some of the troops. In two centres in Cairo and Alexandria alone over 20,000 soldiers attended each station every month.[81] Through these medical interventions the sexual encounter itself was highly regulated. Realising that male sexual appetites could not be curtailed, the authorities tried to ensure that they would be satisfied in a safe way. This meant that the soldier was expected to conduct himself carefully, even at the most intimate moment.

Disciplining the body in active service

It was one thing for the army to provide better sanitary arrangements and medical facilities. In order to be effective, these measures still had to be implemented among troops on the ground. In some instances this meant that the formal codes of discipline that had been established during training were transferred to the field of operations. For example, soldiers continued to be given regular check-ups so that any problems could be identified and remedied. A report on troops in the Mediterranean noted in 1943 that 'health inspections of *all* men in a unit must be carried out efficiently. If V.D., Lice or Scabies exist in a unit inspections will be continued until the unit is clean and healthy.'[82] Known as FFIs, or 'Free From Infection', these medical assessments of soldiers' bodies were highly meticulous even in theatres abroad. Private Henry Butterworth, who served with the 1st Battalion Border Regiment in Egypt, described the scene:

> You would parade in front of the [medical] officer with your trousers loose and your shirt loose. You opened your shirt and let him look under your armpits for lice. You then dropped your trousers and he inspects your nether regions for any other lice. You turn round, bend down for any other lice in the crease of your backside where lice get, you come to attention, put your drawers up and march off, next man in. It must have been a wonderful sight.[83]

Even though this ritual was designed to identify disease or infection, it was still highly militaristic. Each soldier was paraded in front of the medical officer before a sequence of movements and gestures was performed in order to allow for a thorough scrutiny of his body. Once completed, the soldier would then stand to attention and march off. This routine was about more than just about health; the men were expected to obey the commands of the medical inspector.

Treatments and preventions were also compulsory. Private Charles Bennett, who was 'deloused' in Italy, explained that 'you went through the de-lousing process. Showers and then all your kit was stowed. Your clothes were stowed and if you had body hair, all your body hair shaved off. That was the only way you could get rid of it.'[84] Again, these procedures were often highly militaristic. In areas where malaria was a threat, the army carried out quinine and mepacrine parades. Platoon by platoon, men were marched out. Each individual was called forward and given a tablet, which he then had to swallow under the careful watch of the unit commander or medical officer. This was particularly important with quinine, which had a foul taste and was often disliked by the men. Officer James Ford, who served in Hong Kong, explained that 'my sergeant dished it out as I checked through the platoon register and ticked the names off as they came up…Some of the troops used to wait on to see me and the platoon sergeant getting our dose at the end, to make sure that we suffered too, cos it was horrible stuff the quinine.'[85]

As had been the case during training, soldiers who did not adhere to official codes of behaviour also continued to be punished. Private Eric Murray described a fellow recruit in France who 'wouldn't get bathed. He was dirty and we took him down to the showers and, you know them big brooms? Bloody scrubbed him in this washing place in all, that lad. I don't know if he got put on a charge, but you had that kind of thing.'[86] Concealment of venereal disease in particular was a serious offence. Cowell explained that 'it possibly prevents a complete cure, and is a waste of manpower and hospital space.'[87] Penalties included the loss of pay while undergoing treatment in hospital, as well as the loss of any wartime and special proficiency pay. NCOs were liable to lose their acting rank and tradesmen were stripped of their trade rating.[88] Proposing further punishment, an army psychiatrist in North Africa and Italy, Lieutenant-Colonel S.A Mackeith, suggested that no man with venereal disease should be allowed to return home until cured. This, he stated, 'would exercise a powerful effect, especially if adequately publicised.'[89] Even men who submitted to treatment were not free from stigma. Private George Percy recalled that soldiers sent to hospital with venereal disease in Egypt 'had to wear a red tie and that's how everyone knew you'd got VD'.[90] Upon return to active duty, offenders in John Emerson's platoon in Calcutta were punished further still. He recalled that 'we used to see them every evening. They were out with the defaulters in full battle dress, full pack and everything, and they were drilled like hell for an hour, in other words, one heck of a punishment. That went on for about a month.'[91] This public display thus served as both a form of humiliation and a visible warning to other men.

These sorts of punitive measures could, however, have limited effect in the field of active service, where the strict monitoring and surveillance of the soldier's body was not always possible. In theatres abroad, men were often widely dispersed across large areas and were not always under the immediate control of military superiors. In this environment, the army came to rely more on strategies of self-discipline in order to achieve its objectives. In the War Office monograph *Discipline*, Brigadier A.B. MacPherson emphasised how important it was that field troops exercise independent judgement. He stated that, 'Although the principle of unhesitating obedience to orders given by a superior must be firmly maintained, intelligent independence of action in certain conditions must be accepted (and encouraged in training), particularly in conditions where control is difficult, e.g. jungle warfare.'[92] So, while training had gone some way to instilling self-regulation, the military authorities in the field continued to reinforce this quality in the men. In order to do so they focused less on compulsion and more on informal, hegemonic strategies, which were better suited to the conditions of active service. Through a campaign of health education, soldiers were encouraged to look after their own health and fitness and were taught the methods by which to do so. The *Journal of the Royal Army Medical Corps* explained that 'To train a recruit in the techniques of combat is a relatively simple matter, consisting as it does of indoctrination, introduction to weapons and exercise in their use. But to educate such a man so that he may flourish biologically under adverse environmental conditions is exceedingly difficult. Yet this is what must be done for, otherwise, he will quickly eliminate himself by falling sick with preventable disease.'[93] Training in welfare, hygiene and sanitation began back in Britain at the Army School of Hygiene, which was established in Mychett in 1939.[94] Men who passed through the school took their expertise to theatres abroad. In 1942 the commandant of the school stated that 'we are training the soldier to look after himself under any conceivable conditions of climate, anywhere from the Tropics to the Arctic.'[95] A Middle East School of Hygiene was also set up to offer courses of lectures and demonstrations to medical officers, officers and the rank and file. The subjects taught covered all aspects of hygiene, field sanitation and water control. Nearly 200 individuals a month passed through these courses. Richmond and Gear explained that these men returned to their units as 'disciples of the gospel of hygiene'.[96] Courses and lectures were also run at regimental level. According to Major Richards, there were 'travelling circuses' consisting of 'teams of lecturers, model teaching material and many other methods'.[97] A typical example of this was in the Middle East, where demonstration grounds were prepared and full-scale models of all types of sanitary apparatus were shown to the troops.[98] It was also obligatory for hygiene and medical officers to give lectures on issues such as malaria control and prevention of venereal disease. In his memoir, soldier Ben Coutts described a lecture given by the unit medical officer while aboard ship to Egypt. He told the men, 'some of you would put your willies where I wouldn't put my stick!'[99] Private Alexander Frederick served in France. He recalled in an interview that 'the VD warnings frightened the life out of you, frightened the life out of you. They used to show you pictures of a man whose genitals had become infected with gonorrhoea and VD and they used to turn you off, turned you off your food actually.' This was enough to convince Alexander. Horrified by the physical effects of venereal disease, he decided that 'it was best not to risk getting intimate with any women.'[100]

Propaganda was an equally important tool used by the army to persuade men to take responsibility for their own health. Lees highlighted its significance to the prevention of venereal disease:

> This is allied to education but hopes to appeal to produce its effects by different appeal – usually to the emotions, rather than to reason – it hopes to appeal to chivalry, zeal for the cause, pride of self and patriotism, religious motives etc. If cleverly done by an expert it can be invaluable, for it will reach the man to whom reason does not appeal – the rather stupid sensual fellow who indulges most of his appetites – and who is the type most commonly infected.[101]

The Army Bureau of Current Affairs produced fortnightly publications, which covered topics such as 'Social insurance', 'The population problem' and 'The Naples typhus epidemic'. Men were shown short films and humorous cartoons on subjects such as general personal hygiene, prevention of scabies, care of the feet, field sanitation, tropical hygiene, prevention of malaria, prevention of venereal disease and prevention of typhus fever. These were followed by a lecture given by the unit medical officer or a special medical lecturer.[102] In the Middle East leaflets dealing with subjects such as fly control, typhus, malaria and venereal diseases were issued and hygiene education films such as Walt Disney's *Mosquitoes and Malaria* were shown.[103] A 'peppy' publication entitled *Army Illustrated Magazine* was also distributed among troops. According to Richmond and Gear:

> This had a form similar to such popular magazines as 'Lilliput', and 'Men Only', and its presentation of material followed lines suggested by

psychologists. All Army subjects were reviewed in it, such as armoured fighting vehicles, mountain warfare, combined operations, supply systems, etc. These were treated in a simple, bright, concise way and were profusely illustrated. The magazine was eagerly included in the machinery for Army health education. Articles, cartoons, quizzes, covering general health training, physical fitness, fly and mosquito control, and first aid, were published during 1943 and gained much popularity.[104]

Numerous healthcare posters were also produced throughout the war. In North Africa Cowell enlisted the services of London art teacher Stacey Hopper to produce a range of illustrations publicising health issues to troops. These included an image of two mosquitos having tea, with one saying to the other, 'But my dear! You must try the troops', followed by the caption 'Don't give them the chance – mosquitos mean malaria!'[105] Deployed later to Italy, Hopper also created posters designed to encourage sexual abstinence. These included an image of busty woman that warned 'Just another "Claptrap". You can ruin your future with VD!'[106] Such images once again targeted the female body as a particularly infectious site. Recalling his experiences in Calcutta, George MacDonald Fraser also described one poster that 'showed a statuesque blonde, surrounded by leering Japanese, with the caption: "Is this the face that loved a thousand Nips?"'[107] In this instance, it appears that the authorities hoped to appeal to men's sense of loyalty and patriotism, as troops were warned to avoid a dangerous female body that had been defiled by the enemy.

The released body

Released from the constraints of formal military discipline that had characterised life in barracks, men on active service had more authority over their bodies than ever before. For some, the self-discipline that had been instilled during training did find practical application in the field, even in the harshest of conditions. Charles Bennett fought at Monte Cassino in Italy in 1944 and recalled that, 'well, you did your best. I mean, it had become part of your training to wash and shave, but a bit difficult when it's ice and snow, you've only got icy water. And the biggest thing was, of course, you couldn't get baths or anything like that.'[108] Others were, however, less enthusiastic about caring for their own health and wellbeing. In the Middle East, Richmond and Gear noted that attitudes to the importance of disease and the need for high standards of health and cleanliness 'varied considerably'. While some units and individuals had 'the right principles', others 'were simply contemptuous of real

Figure 3 'Just another "Claptrap". You can ruin your future with – V.D.!' Poster designed by Stacey Hopper to warn Allied troops in Italy about the dangers of venereal disease, 1943–44

soldiers bothering themselves with anything so childish or mundane as killing flies, avoiding mosquitoes or being particular as to the state of the cleanliness of their camps, kitchens or their persons'.[109] Despite the army's efforts to educate men about the importance of personal health and hygiene, it seems that some were reluctant to embrace new practices, regarding such matters as unimportant, juvenile and even unsoldierly.

All of the army's warnings, education and propaganda also did little to curb the sexual appetites of British men serving overseas. In Italy, for example, regardless of appeals for abstinence, hospital admissions for venereal disease increased from 27 per 1,000 in 1943 to 64 per 1,000 in 1945.[110] In north-west Europe, after the invasion of Normandy, cases rose from 2 men per 1,000 to 26 per 1,000. In the Middle East, the number of incidences of venereal disease stayed roughly the same between 1942 and 1945 at around 28 men per 1,000 troops.[111] George MacDonald Fraser, who had the task of raiding the Cairo Berka, described what little effect the policy of placing it 'out of bounds' had on men's sexual ambitions:

> As a harassed orderly officer in North Africa I had to raid more brothels, endure the screaming protests of more furious harlots, and see more frustrated amorists thrown into the paddy-wagon than I care to count, and I remember the blue and Khaki queues outside the bawdy-housed of the Cairo Birkah, the more impatient customers already in their shirt-tails with their trousers neatly folded over their arms.[112]

According to army psychiatrist Lieutenant-Colonel MacKeith, this kind of behaviour was a direct response to the conditions of active service and was even an outright act of defiance, occurring when the soldier was frustrated with his military superiors. He suggested that 'promiscuity occurs very much more when the soldier is either "strung-up" or "tense" after being in action or is discontented, disgruntled, critical of his officers and N.C.O.'s and is generally "browned-off". In such circumstances, men tend to lapse into a short-sighted, despairing and childish state of mind.'[113] Soldiers' testimonies suggest, however, that the reasons that men chose to be promiscuous were more varied and complex. Some did link their impulses specifically to the conditions of warfare. Official rules and regulations could, for instance, mean little to men who felt that they could be killed at any moment. Henry Butterworth, who was a frequent visitor to the Cairo brothels, felt this way:

> We were ordered not to go to any place in Cairo of ill-repute. You know what I mean [laughs]. As young soldiers we were raring to go, and if you got caught it was a prison sentence, no trouble at all…All the houses of ill-

repute, and there were hundreds of them, all had great big notice outside, 'out of bounds', so thy advertised the fact [laughs]. This was the life. You'd only got three weeks to live. That's how we looked at it. We couldn't care less.[114]

Believing that their days were numbered, Henry and his friends acted with little fear of repercussion. Indeed, it appears that the prospect of death heightened their sexual ambitions, as they became less focused on a long-term future and more on enjoying themselves while they could.

Other soldiers did not, however, frame their behaviours within the conditions of war. For young men away from home, often for the first time, service abroad simply introduced new sexual motives and opportunities, and many made the most of any chances to indulge. George Percy, who visited the Berka after it had been closed to British troops, explained that 'to me it was an adventure. I mean for a young lad…I really enjoyed it.'[115] Men deployed to far off and exotic countries also welcomed the opportunity to sleep with women of different races and nationalities. They were often surprised and delighted at how different these women were to those they had left behind. Foremost in Eric Murray's memories of a brothel in Haifa was 'a coloured girl. She was bonnie and smart looking and they were all bloody queuing up for this one in there.'[116] Officers were just as prone to temptation. Lieutenant John Emerson, who described himself as 'a young and healthy man with the correct idea on life' and 'a great believer in sex', visited a brothel in Calcutta that was 'stocked with European women from every country you could think of'. He met 'a charming Viennese woman among the group there and had a wonderful evening.'[117]

As had been the case during training, sexual exploits were also not just about physical gratification. While high-ranking medical officers may have constructed notions of military pride and honour around qualities of abstinence and self-control, as we have seen, a preoccupation with sex and the ardent pursuit of women were prized masculine traits among the men. George MacDonald Fraser highlighted why men might want to appear promiscuous when he described the actions of one of his fellow recruits while on leave in Calcutta:

> Forster was an interesting case. On our first night in Calcutta he spruced himself up with Lifebuoy, flourished the prophylactic kit which he had drawn from the M.O.'s office, admired himself in the mirror, sketched out a programme of debauchery which would have frightened Caligula, and strode forth like Ferdinand the Bull. Three hours later he was back, full of gloating accounts of his sexual heroics, and unaware that in the interval

Grandarse and I had been sitting three rows behind him in the Lighthouse cinema, watching Laurel and Hardy.[118]

Forster therefore engaged in a performance in order to assert a particular self-identity associated with virility and heterosexuality. He went to great efforts to project this impression of himself to the group. Clearly, it was the opinions of his fellow soldiers that mattered most to him.

Soldiers also ignored official advice regarding sexual behaviour for more practical or mundane reasons. Some chose to practise unsafe sex or not to treat themselves properly because they were hot, tired, drunk, could not be bothered, or did not know how to use the prophylactics. Others preferred to use their condoms to waterproof wallets or watches.[119] In the Middle East Lees reported, 'I find that soldiers will not use antiseptics, etc., before coitus.'[120] He also believed that men were not using condoms, even though they were supplied free, because of 'carelessness, laziness, drunkenness, and diminished pleasure in the sexual act'.[121] Director of Army Medical Services, Lieutenant-Colonel Alexander Hood likewise noted that 'it is, in my experience, quite exceptional to find that the packets have been used in the way they are intended.'[122] He questioned whether troops were carrying out self-disinfection properly in preventative ablution centres and argued that more stringent medical supervision was necessary. Hood stated, 'a large number of the men who use the room are content to do so in a perfunctory manner in order to obtain a ticket which may absolve them from the loss of privileges later. There are many factors accountable for this: fatigue, alcohol, hot weather, a desire to get back to barracks etc., all play their part.'[123] In the opinions of these medical staff, soldiers were simply not capable of administering their own prophylactics or could not be trusted to do so.

As well as getting drunk, soldiers on active service also continued to malinger, go absent or self-inflict wounds. All of these offences were punishable by court martial. Men found guilty faced loss of pay and a period of detention, ranging from ten days for drunkenness to between three and five years for desertion.[124] Convictions increased throughout the war, as Table 2 shows.

While the official statistics do not reveal the full extent of these behaviours, they do, nevertheless, suggest a correlation between these offences and the intensity of fighting.[125] Over half of convictions for all offences took place between 1944 and 1945, the year of the invasion of Normandy. Of the two biggest crimes, desertion and absence without leave, over 60 per cent of convictions occurred in this period, as did 80

Table 2 Court martial convictions, British other ranks overseas, 1 September 1939–31 August 1945

Offences	1939–40	1940–41	1941–42	1942–43	1943–44	1944–45	Total
Cowardice (including malingering)	0	3	6	23	41	70	143
Desertion	23	70	647	1,083	1,827	8,425	12,075
Absence without leave	339	284	1,113	2,101	2,992	8,085	14,914
Self-inflicted wound	5	7	18	8	19	208	265
Drunkenness	409	153	70	630	714	926	2,902
Total	776	517	1,854	3,845	5,593	17,714	30,299

Source: TNA WO277/7, Comprehensive summary of court-martial convictions (British other ranks), 1 December 1939–31 August 1945.

per cent of convictions for self-inflicted wounds. Offences like these were also more common in the infantry, whose role MacPherson described as 'dangerous, arduous, unspectacular and unrewarding'.[126] In the Second Army, 21st Army Group, serving in Normandy, for example, of a total of 388 suspected cases of self-inflicted wounds, sent to No. 110 British General Hospital in 1944, 279 were infantrymen, a figure out of all proportion to their numbers.[127] In a study of 2,000 desertion cases in the British Army of the Rhine, 1,770 were infantrymen and 47 were in the Royal Artillery. Again, these figures were out of all proportion to the relative strength of these two arms of the service. The majority of cases occurred among soldiers who had been in active service for between two and three years, which suggests that these individuals may have been growing increasingly desperate.[128]

Soldiers' personal accounts again tell us more about why men chose to get drunk, harm themselves or leave their units. Infantryman Ronald Sherlaw served in Italy between 1943 and 1944. He believed that alcohol was used as a coping mechanism for men who had been in active service for extended periods. He explained that 'Battle weariness evidenced itself out of the line rather than in the line. People got involved in excesses of one kind or another. They started being silly over drink and things like that.' Ronald stated that this was more common among men who had never been injured because, 'if you were never wounded, you never got a rest.'[129]

Malingering or self-inflicting wounds was therefore one option available to men who wanted to remove their bodies from the combat environment, even just for a brief spell in hospital. In his memoir, infantryman Rex Wingfield described the advice given to him by fellow patients on how to feign illness, as he recovered from the effects of the cold in a regimental aid post in Holland. These included putting toothpaste under the tongue to produce a high temperature when a thermometer was placed in the mouth, chewing cordite, which created a fast heart rate, and swallowing cotton wool balls, which showed up like duodenal ulcers on an x-ray.[130] Indeed, the desire to malinger was played upon by the enemy. In Italy the Germans distributed hundreds of 'Malingerers Guides' to soldiers. Disguised as matchbooks, these contained minutely printed instructions on how to simulate almost every disease, ranging from minor troubles such as inflammation of the foot to serious conditions like heart disease and tuberculosis. For example, in order to feign a sore throat, men were advised to 'Take a silver nitrate stick and rub a small area of your tonsils with the tip of the stick' before drinking yellow mustard or ginger in water in order to produce 'a burning sensation on the tip of your tongue'. The instructions concluded that the 'treatment is absolutely harmless. All of the resulting symptoms disappear completely in a few days. Remember: Better a few weeks ill than all your life dead.'[131]

In order to harm their bodies, some men refused to take proper anti-malarial precautions. Infantryman Gerald Barnett recalled that in Italy 'some people tried to get malaria if they could, by letting mosquitos bite them.'[132] Others tried to get a 'Blighty' wound, which was bad enough to get them out of the line but not so serious as to be life threatening.[133] Charles Bennett remembered a fellow soldier in Italy who 'shot himself in the foot cos he couldn't stand the shelling.'[134] On active service in India in 1944, William Cornell encountered a fellow soldier who, even before going into combat, shot himself in the foot. When William told him, 'you're in trouble now,' he replied, 'I know but I cannot face it...I couldn't go any further.'[135] Such was the extent that men would go to in order to escape life on the front line and perhaps the possibility of more serious bodily injury. By inflicting wounds on themselves, soldiers could control the damage to their bodies, rather than taking their chances with the enemy.

Civilians into soldiers

Soldiers' testimonies also reveal, however, that men did not just commit these sorts of offences in moments of crisis. Like sex, drunkenness was often opportunistic and the result of unfamiliarity with local liquor. British servicemen in overseas countries tasted drinks like wine, cognac, Bols and Dutch gin for the first time.[136] Inexperienced drinkers were unaware of the potential effects of these strong spirits. Driver Gordon Gent served in Algiers and recalled that 'the men thought you could drink these wines just like beer in England, but of course you couldn't drink it like beer. They used to get in the most holy states with it. They would go mad with it.'[137] Tank driver John Buchanan was introduced to Calvados while serving in France. He remembered that 'it went right down to your toes and came stotting back up again.' John recalled one particular drunken night in camp when 'one of the boys went mad. We had to sit on him. He was screaming. It was this Calvados that set him off, you see. Twenty-four hours later he came to. It was a worry for us because we weren't supposed to be drunk. We must have been out of action for a couple of days...The hangover from Calvados is something else.'[138] Charles Jordan likewise suffered after a night drinking the local beer in Hong Kong. A gunner in the 18th Coastal Regiment, Royal Artillery, Charles was put on a charge when he returned to camp.

> I only went on what might call a binge on one occasion and that was at Christmastime. It didn't end very well for me because the first time I ended up on an army charge. I went into Hong Kong with several companions. We went round the hotels and we were drinking beer. The Chinese beer is nothing like you get at home. It was called the Hong Kong Brewery. I always remember that. It wasn't brewed. It was chemical beer, which gave you a very bad hangover and sickness feeling. That was the one and only lesson I had drinking out there as regards that beverage. I knew I'd had enough and I decided to return to camp. I wasn't drunk but I knew I was going to be very bad.[139]

Men also deserted for all sorts of reasons. In 1945 Private Sydney Cook of the Loyal Regiment, British Expeditionary Force, was sentenced to five years' penal servitude after being found guilty of desertion 'for the purpose of internment for the duration of the war'. Sydney had tried, but failed, to avoid conscription on conscientious grounds. Recruited and deployed to France in January 1940, he remained so morally opposed to the war that he crossed the border from France into Belgium, where he knew he would be captured and imprisoned by the enemy. He was held at a prisoner of war camp in Brussels until June 1945.[140] In a

Active service

private letter written while in captivity to his commanding officer, Cook asserted his objection to 'the taking of human life', stating, 'I will gladly suffer death rather than raise a hand against those that would kill me.'[141] Other men deserted to escape the everyday discomforts of life in active service. In 1942 Lieutenant J. Speight, of the 4th Queen's Own Hussars, Tank Delivery Regiment, was sentenced to five years' imprisonment for deserting when on active service in a forward area. Having left his regiment while it was stationed at Hammimeat in Egypt, he was arrested in Cairo a month later. In a statement to the court martial he explained that 'I do not wish to offer any excuse for my behaviour except that I had been bothered a great deal by desert sores and the flies got on my nerves a great deal.'[142] Indeed, some men went absent simply to have a good meal or to get a good wash. During the Battle of Monte Cassino in 1943 several of the men in Kenneth Bond's unit left their positions:

> We were in this big ravine sort of place, a wadi, as it was called. We were under canvas sort of tents. You know, it was very makeshift, very, and we were getting blankets soaked and we weren't being able to sort of clean or wash properly, no bath or anything like that. I know they used to say at least three a night used to go back to what we called B Echelon, that was the base of your unit, and give themselves up…They said, 'We're not deserting. We just want to go back and have a wash and brush up,' and that's not what we were taught to do.[143]

Simply existing in the combat environment was enough to make these men abandon their responsibilities, albeit temporarily. They did not intend to abscond from the army but simply wanted a chance to refresh their dirty and weary bodies. Nevertheless, they were charged and convicted of desertion and were given two-year suspended sentences.[144]

In the field of active service it was also more likely that some infractions would be tolerated by those in charge. More concerned with fighting the enemy, officers often did not interfere with their men's behaviours as long as they did not undermine efficiency or morale. Roy Bolton served in France and recalled that:

> You were expected to be soldier-like at all times. I think on about the very first or second day there, after our arrival in this field, the RSM pulled us, told us, got us all together, those that were on duty, and gave us a pep talk about the importance of keeping your boots polished and your equipment clean, and wearing it at all times, being ready for instant action. He concluded his remarks by saying, 'You think that now you're in action you can do what you like, but you're still soldiers, you'll still do what I tell you to do and you will be properly dressed at all times,' and so on. I would add

in parenthesis that that didn't last very long and later on we were left much to ourselves as long as we did our work.[145]

Richmond and Gear highlighted the lax attitude of some officers towards malaria prevention in the Middle East. They expressed the belief that no amount of 'propaganda, exhortation or training' would make any army malaria conscious. A battalion on dispersed guard duty along the Syrian border, for instance, 'in spite of over a year's experience in the Middle East, neglected such precautions as checking that all men were wearing slacks, applying mosquito cream, and using sleeping nets, with the result that scores of cases occurred in a few weeks.'[146] Mark Harrison has also shown that preventative measures for the suppression of malaria in Italy were never enforced by officers, which led to over 20,000 British troops being admitted to hospital for the disease during the invasion of Sicily in 1943.[147] The neglect of bodily regimes in these instances appears to represent the relaxation of discipline rather than resistance on the part of the men.

Unlike senior staff, officers on the ground, who were better acquainted with the needs of their men, also took a relaxed view of soldiers getting drunk or being promiscuous. Alex Gilchrist, an officer in the 2nd Fife and Forfar Yeomanry served in Ypres in early 1945. He remembered that the men were always 'going out drinking too much et cetera and looking for women all the time. I mean, if there was a woman to be had anywhere at all, they were after them, but I mean they were all good chaps to me. They did what I wanted them to do and that was all I asked of them.'[148] When four of the men in Ernest Lanning's platoon swam a river one night to visit an Italian Army brothel in Abyssinia, Ernest's attitude when they returned was 'jolly good show, chaps'.[149] Officer John Emerson disagreed with the placing of brothels out of bounds to the rank and file in Calcutta. He believed that it was better to provide regulated establishments for the troops, 'because you can control it'.[150] While he certainly did not share these views on sexual promiscuity, even MacKeith thought that, in moderation, alcohol could be an effective stress reliever:

> During off duty periods drink can subserve good functions. It helps the party spirit, and in soldiers' minds is closely associated with jollification. Over and beyond this, for men who after battle, though superficially all right, are 'strung-up' and 'tense' in the back of their minds, drink can provide an effective and natural method of 'drowning their sorrows' – far less harmful than the other method which consists in promiscuous intercourse. In fact in such circumstances drink *can* be a harmless substitute for promiscuity. Under suitable circumstances, it may well as such be unofficially condoned.[151]

Active service

Away from the strict segregation of ranks that had occurred in barracks, officers themselves drank to excess during service abroad and encouraged their men to do so. During a night out in Tel Aviv, Private Eric Murray was introduced to the alcoholic drink arak by his corporal, a man named Lees:

> I had some and I'm drinking and I could feel it, this, this is getting a hold of me. So I just went out. I went down on the seafront. I like to drink stuff when I know what effect it's going to have, but I could feel this going on me and I said, 'I'm not having any more.' So I went down on the seafront and I stopped down there for maybes three-quarters of an hour, an hour. When I came back they were underneath the table [laughs]. So, so I had to get them back.[152]

Newcomers to Gerald Barnett's platoon in North Africa were invited to play the infamous drinking game 'Colonel Puff'. Participants recited a poem, followed by a sequence of finger and toe tapping, before taking a drink. If a mistake was made, the man would have to empty his glass. According to Gerald, 'the number of times a man could repeat the sequence determined his social standing.'[153] As such, this performance constituted an initiation, or a rite of passage, into the group.

Even where certain behaviours were not condoned, there were also practical issues to be considered when deciding whether to punish a man for his transgressions. It is likely, for example, that the official statistics significantly underestimate cases of self-inflicted wounds, which represented less than 1 per cent of all court martial convictions among troops overseas.[154] David French suggests that that this is because military superiors were often reluctant to report men for causing their own injuries, either from compassion or because prosecution might reflect badly upon their units.[155] Punishing a soldier with imprisonment could also seem futile when the armed services were so desperate for manpower. Medical orderly James Bell was instructed by his chief medical officer in Shanghai not to report men who contracted venereal disease and had not submitted to disinfection procedures. Instead, he was told to give them a slip, tell them to come back in two days, and then treat them. This way it appeared that the men had followed official regulations. According to James, it 'seemed to make more sense than 28 days in the "glasshouse".'[156]

A sense that the rules relating to the body were more relaxed in the context of active service was also evident when punishments were inflicted for misdemeanours. A sentence of between three and five years' imprisonment for desertion was, for instance, impracticable during wartime because it meant the removal of the offender from the front line and required guards to watch him.[157] This punishment was also no deterrent for men who preferred the comparative safety of a detention barracks to the hardships of war.[158] Until 1930 desertion in active service had been a crime punishable by death.[159] During the First World War, 264 soldiers had suffered this fate, as well as many others who were unofficially shot while running away.[160] The abolition of the death penalty for desertion during the interwar period may have made it a more attractive option for men who wanted to escape combat life. Indeed, the percentage of court martial convictions for desertion and absence without leave was greater among troops overseas during the Second World War. While 25 per cent of all convictions had been for these offences between 1914 and 1918, this figure increased to 40 per cent between 1939 and 1945.[161]

Faced with the impracticalities of detaining men who absconded, soon after the outbreak of the war the army decided to release, either by suspension or remission of sentence, all soldiers whose unexpired sentences did not exceed three months. In the 21st Army Group in Europe a special Review of Sentence Board was set up by 2nd Army headquarters to interview and select men who should be quickly returned to their units from all cases that had been sentenced for desertion or kindred offences after three months had been served. Of the 596 cases initially reviewed, 435 were returned to the front line.[162] Similar bodies were set up both in the 21st Army Group and in other commands overseas. Of the first 2,000 cases considered, 75 per cent were released under suspended sentence after the first interview.[163] Fearing that this new system was too lax, several army commanders requested that the government reintroduce the death penalty for desertion on active service.[164] In a letter to the War Office in April 1942 General C.J.E. Auchinleck, Commander-in-Chief of the Middle East Forces, talked about the 291 convictions that had taken place under his command in the previous year, stating, 'I have no doubt that had the enforcement of the death sentence been within my discretion, the knowledge of this fact would have proved a statutory deterrent in a number of cases in which the worst example was set by men to whom the alternative of prison to the hardships of battle conveyed neither fear nor stigma.'[165] Auchinleck also pointed out that those men whose sentences had been commuted often went on to repeat their crimes. In another telegram he described 'typical bad cases', such as a guardsman who had been sentenced to two years' imprisonment in February for deserting with a truck, who had served four months of his sentence with the remainder suspended and re-joined his battalion in June. Ten days later, the man deserted again, taking a truck containing 60 gallons of water and twenty-three other men.[166] Despite such requests the British government refused to reintroduce the death penalty for desertion, due to the political climate at home. Reintroduction of the death penalty would mean publicly disclosing the army's apparent poor morale, and thereby provide ample opportunity for enemy propaganda. The authorities would also have to be able to distinguish genuine cases of mental breakdown from malingerers so as to execute only those men who preferred prison to the front line.[167] Thus, for high-ranking army commanders the deserter's body was symbolic as something to be punished in order to uphold military tradition and morale. However, for government officials, it was a practical concern and public symbol that was critical to the war effort on the home front.

Conclusion

The combat environment placed new demands upon the soldier's body in certain areas of his day-to-day existence. While training had involved relatively brief tests of men's skills, fitness and endurance, they now faced long periods of exposure to climatic extremes, hostile landscapes and an array of new diseases. The body therefore remained a crucial concern for the military and medical authorities once men had been deployed overseas. As Harrison suggests, a particular 'medical consciousness' was evident among British forces in the field, where officers and commanders strove to keep their men free from disease.[168] What these efforts also represent is a particular body culture, which can be linked to the colonial past. The authorities clearly recognised the role played by the physical environment in matters of human health. In some instances the body was perceived to be open and in flux with the environment so the army adopted various techniques that were aimed towards physiological adaptation. These included programmes of vaccination, immunisation, and acclimatisation, which were designed to boost the body's natural powers of defence. The body was also, however, conceived as limited in its adaptability so needed protection from the surrounding environment. As such, the army adopted a range of 'defensive mechanisms', including different types of clothing and mosquito nets. Much of this work was carried out by the Army Hygiene Service, which provided clean water, ablution facilities and disinfection and disinfestation. Through these mechanisms the army sought to negate the harmful physiological effects of the environment by manipulating the landscape itself.

It was not just the environment that posed new risks to the soldier's body. British troops were also under threat from dangerous other bodies, namely enemy troops and native populations, whose poor sanitary habits were considered to be a root cause of disease. The indigenous female body was particularly isolated as a biohazard, as a potent carrier of venereal disease. Again, the authorities adopted a two-pronged strategy in order to protect their men. When possible, soldiers were separated from native towns and communities and were banned from going to brothels. However, the army also focused on reforming indigenous bodies by implementing sanitation and disinfection programmes, arresting and imprisoning prostitutes and establishing regulated brothels.

Yet, while it is clear that the army's management of the body continued in theatres abroad, what did change was the nature of the control exerted upon it. The conditions of active service could make monitoring men more difficult, so the army adopted more informal, hegemonic strategies in order to try and keep troops free from disease. Through a campaign of health education and propaganda, men were encouraged to look after their own bodies and were taught the practical skills by which to do so. These methods were sometimes successful, persuading troops to wash and shave or discouraging them from having sex. Nevertheless, many other soldiers chose to ignore the advice they were given. Free from the formal codes of monitoring and surveillance that had been imposed during training, men chose to be sexually promiscuous, get drunk or absent themselves from duty. Some did so as a direct response to the harsh realities of life in active service. Soldiers who feared death, were constantly cold, wet, tired or battle weary, easily succumbed to temptation. Others simply made the most of the opportunities available to them, particularly the chance to sleep with foreign women. These behaviours must also be seen as evidence of the widening of disciplinary parameters or of a resistance that was more than ever before legitimised by those in charge. In countries abroad officers did not report men for drunkenness and promiscuity as long as it did not threaten the task at hand. Some even got drunk with their men. The army also faced practical dilemmas when it came to charging men for these indiscretions. Imprisonment was, for example, a waste of manpower and resources, so the army adopted a policy of releasing offenders before their sentences were up. Knowing this, it was perhaps easier for men to break the rules.

Active service

While active service brought about new physical challenges and new methods to solve them, with this came a level of personal freedom that few soldiers had experienced before. For many, however, the toughest challenges to their bodies would come when they faced being injured or killed.

Notes

1. *Strength and Casualties of the Armed Forces* (Cmd. 6832), p. 4.
2. War Office, *Handbook of Military Hygiene* (London: HMSO, 1943), p. 4.
3. J. Ellis, *The Sharp End: The Fighting Man in World War II* (London: Aurum Press, 2009), p. 181.
4. Harrison, *Medicine and Victory*.
5. Harrison, *Medicine and Victory*, p. 91.
6. Hockey, 'Head down, bergen on', 150.
7. Hockey, 'Head down, bergen on', 168.
8. War Office, *Handbook of Military Hygiene*, p. 11.
9. E.T. Burje, *Tropical Tips for Troops in the Tropics or How to Keep Fit in the Tropics* (London: Heinemann, 1942), p. 2.
10. See M. Harrison, *Climates and Constitutions: Health, Race, Environment and British Imperialism in India, 1600–1850* (Oxford: Oxford University Press, 1999); E.M. Collingham, *Imperial Bodies: The Physical Experience of the Raj, c.1800–1947* (Cambridge: Polity, 2001).
11. War Office, *Handbook of Military Hygiene*, p. 11.
12. 'The army medical services: wartime activities and developments', *Journal of the Royal Army Medical Corps*, 82:6 (1944), 256.
13. TNA FD1/6383, A.D.M.S.2/7, Acclimatization of troops, 6 November 1944, p. 1.
14. Burje, *Tropical Tips for Troops*, pp. 17, 19.
15. BBC WW2 People's War Archive, A7747725 E. Owen Proctor, 13 December 2005, www.bbc.co.uk/history/ww2peopleswar/stories/25/a7747725.shtml (accessed November 2013).
16. BBC WW2 People's War Archive, A2772489, Harry Blood, 23 June 2004, www.bbc.co.uk/history/ww2peopleswar/stories/89/a2772489.shtml (accessed November 2013).
17. TNA FD1/6383, A.D.M.S.7/1, Acclimatization of troops, 6 November 1944, p. 1.
18. 'Hygiene, morale and desert victory', *British Medical Journal*, 1944 (18 March 1944), 397; M.A. Osbourne, 'Acclimatizing the world: a history of the paradigmatic colonial science', *Osiris*, 2nd series 15: Nature and Empire (2000), 135–51.
19. Col. A.E. Richmond and Lt.-Col. H.S. Gear, 'The health of the Middle East force, 1942–1943', *Journal of the Royal Army Medical Corps*, 85:1 (1945), 7.

20 War Office, *Handbook of Military Hygiene*, p. 24.
21 'The army medical services: wartime activities and developments', 260.
22 War Office, *Handbook of Military Hygiene*, p. 24.
23 Col. J.S.K. Boyd, 'Enteric group fevers in prisoners from the Western Desert', *British Medical Journal* (12 June 1943) 719–21.
24 War Office, *Handbook of Military Hygiene*, p. 9.
25 F.A.E. Crew, *The Army Medical Services: Administration, Volume II* (London: HMSO, 1953), p. 69.
26 Crew, *The Army Medical Services: Administration, Volume II*, pp. 72–4.
27 'The army medical services: wartime activities and developments', 256.
28 Burje, *Tropical Tips for Troops*, p. 4.
29 Richmond and Gear, 'The health of the Middle East force', p. 15.
30 Richmond and Gear, 'The health of the Middle East force', p. 9.
31 War Office, *Handbook of Military Hygiene*, p. 15.
32 War Office, *Handbook of Military Hygiene*, p. 8.
33 W.G.V. Balchin, 'United Kingdom geographers in the Second World War: a report', *Geographical Journal* 53:2 (1987), 159–80.
34 TNA WO252/1316, Report by Inter-Service Topographical Department on Southern Italy: climate and medical, 29 June 1943, pp. 8–9.
35 'The army medical services: wartime activities and developments', 256
36 Crew, *The Army Medical Services: Administration, Volume II*, p. 26.
37 Richmond and Gear, 'The health of the Middle East force', 18–21.
38 Richmond and Gear, 'The health of the Middle East force', 13.
39 H. Pozner. 'Medical history of an action', *Journal of the Royal Army Medical Corps* 83:4 (1944), 183–4.
40 'Hygiene, morale and desert victory', *British Medical Journal* (18 March 1944), 398.
41 Lt.-Col. H.S. Gear, 'Hygiene aspects of El Alamein', *British Medical Journal* (18 March 1944), 384.
42 TNA WO252/1316, Report by Inter-Service Topographical Department on Southern Italy: climate and medical, p. 9.
43 Gear, 'Hygiene aspects of the El Alamein victory', 383.
44 MOA TC29, Forces: Men in the Forces 1939–1956, 3/A, Health memoranda for British soldiers in the tropics, June 1943, pp. 4, 12.
45 See, for example, V. Talwar Oldenburg, *The making of Colonial Lucknow* (Oxford: Oxford University Press, 1989).
46 Richmond and Gear, 'The health of the Middle East force', 18.
47 Harrison, *Climates and Constitutions*, p. 22; M. Ramanna, 'Perceptions of sanitation and medicine in Bombay, 1914–1918', in H. Fischer-Tine and M. Mann (eds.), *Colonialism as Civilizing Mission: Cultural Ideology in British India*, (London: Anthem Press, 2004), pp. 205–24.
48 Harrison, *Medicine and Victory*, p. 94.
49 Richmond and Gear, 'The health of the Middle East force', 2.

50 Harrison, 'Sex and the citizen soldier', pp. 234–5.
51 TNA WO222/1302, Lt.-Col. Robert Lees, Methods of prevention of venereal disease, 14 April 1942, p. 1.
52 MOA TC29, Forces: Men in the Forces 1939–1956, 3/A, Health memoranda for British soldiers in the tropics, p. 11.
53 Wellcome Archives, RAMC 466/48, Lt. E.M Cowell, Health notes from the office of the Surgeon AFHQ, 1944, p. 2.
54 Wellcome Archives, RAMC 466/48, Lt. E.M Cowell, Health notes from the office of the Surgeon AFHQ, 1944, p. 2.
55 D.M. Peers, 'Soldiers, surgeons and the campaigns to combat sexually transmitted diseases in Colonial India, 1805–1860', *Medical History* 42:2 (1998), 146.
56 F.B. Smith, 'The Contagious Diseases Acts reconsidered', *Social History of Medicine* 3:2 (1990), 197.
57 L. Bland, '"Guardians of the race" or "vampires upon the nation's health"? Female sexuality and its regulation in early twentieth-century Britain', in E. Whitelegg et al. (eds.), *The Changing Experience of Women* (Oxford: Martin Robertson in association with the Open University, 1982), pp. 375–88.
58 *Summary Report by the Ministry of Health for the period from 1st April, 1941 to 31st March, 1942* (Cmd .6394), p. 6.
59 'New compulsory powers in control of venereal disease', *Lancet* (14 November 1942), 589.
60 R. Davidson, '"Searching for Mary, Glasgow": contact tracing for sexually transmitted diseases in twentieth-century Scotland', *Social History of Medicine* 9:2 (1996), 200.
61 TNA WO177/1, Director of Medical Services, British Expeditionary Force, Medical administrative instructions no. 12, 23 Nov 1939, p. 2.
62 Wellcome Archives, RAMC 466/48, Brig. R. Lees, Col. T. Young and Lt.-Col. D.J. Campbell, Recommendations for the prevention of V.D. amongst Allied forces in the central Mediterranean theatre of war, 1943, p. 2.
63 TNA WO222/1302, Lees, Methods of prevention of venereal diseases, p. 1.
64 J. McCallum, *Journey with a Pistol* (London: Victor Gollanz, 1959), p. 68.
65 IWM SA, 24187, George Percy, reel 7.
66 Richmond and Gear, 'The health of the Middle East force', 28; TNA WO222/1302, Lees, Methods of prevention of venereal diseases, p. 2. For a discussion of the army's practice of placing brothels out of bounds during the First World War, see M. Harrison, 'The British Army and the problem of venereal disease in France and Egypt during the First World War', *Medical History* 39 (1995), 133–58.
67 Harrison, *Medicine and Victory*, p. 104.
68 O. Anderson, 'The growth of Christian militarism in mid-Victorian Britain', *English Historical Review* 86 (1971), 46–72.

69 The practice of placing brothels 'out of bounds' had originated in France and Egypt between during the First World War, as a response to increasing pressure from moral reformers. Harrison, 'The British Army and the problem of venereal disease'.
70 Harrison, 'Sex and the citizen soldier', p. 226.
71 Wellcome Archives, RAMC 466/48, Cowell, Health notes from the office of the Surgeon AFHQ, 1944, p. 2.
72 TNA WO222/1302, Lees, Methods of prevention of venereal diseases, p. 1.
73 Wellcome Archives, RAMC 466/48, Lees, Young and Campbell, Recommendations for the prevention of V.D. amongst Allied forces in the central Mediterranean theatre of war, p. 1.
74 A.N. Gilbert, 'Law and honour among eighteenth-century British Army officers', *Historical Journal*, 19:1 (1976), 76.
75 War Office, *Manual of Military Law*, pp. 442–4.
76 TNA WO222/1302, Lees, Methods of prevention of venereal diseases, p. 2.
77 'Prevention of venereal disease in the army', *Journal of the Royal Army Medical Corps* 81:1 (1943), 36.
78 Wellcome Archives, GC/135/B.1/3, Lt.-Col. S.A. MacKeith, 'Some comments on the V.D. problem in an expeditionary force', 1944, p. 1.
79 Harrison, 'The British Army and the problem of venereal disease', 147–8; P. Levine, *Prostitution, Race and Politics: Policing Venereal Disease in the British Empire* (London: Routledge, 2003), pp. 147–8.
80 Harrison, 'The British Army and the problem of venereal disease', 158; Levine, *Prostitution, Race and Politics*, p. 148–9.
81 TNA WO222/1302, Lees, Methods of prevention of venereal diseases, p. 2.
82 Wellcome Archives, RAMC 466/48, Lees, Young and Campell, Recommendations for the prevention of V.D. amongst Allied forces in the central Mediterranean theatre of war, 1943, p. 2.
83 IWM SA, 23367, Henry Butterworth, reel 2.
84 IWM SA, 13230, Charles Bennett, reel 3.
85 IWM SA, 13128, James Ford, reel 2.
86 IWM SA, 17630, Eric Murray, reel 4.
87 Wellcome Archives, RAMC 466/48, Cowell, Health notes from the office of the Surgeon AFHQ, 1943, p. 2.
88 Wellcome Archives, RAMC 466/48, Cowell, Health notes from the office of the Surgeon AFHQ, 1943, p. 2.
89 Wellcome Archives, GC 135/B.1/3: Lt.-Col. S A MacKeith, Some comments on the VD problem in an expeditionary force, 1944, p. 5.
90 IWM SA, 24187, George Percy, reel 7.
91 IWM SA, 30078, John Emerson, reel 3.
92 TNA WO277/7, Brigadier A.B. MacPherson, Discipline 1939–1945, 1950, p. 21.
93 'The army medical services: wartime activities and developments', 258.

94 Crew, *The Army Medical Services: Administration, Volume I*, p. 51.
95 'Hygiene in the army: modern methods of teaching', *The Times* (11 March 1942), p. 2.
96 Richmond and Gear, 'The health of the Middle East force, 1942–1943', 5.
97 Maj. H.J.A. Richards, 'Health education in the army', *Journal of the Royal Army Medical Corps* 90:4 (1948), 137.
98 Richmond and Gear, 'The health of the Middle East force', 6.
99 B. Coutts, *A Scotsman's War* (Edinburgh: Mercat Press, 1995), p. 22.
100 IWM SA, 19804, Alexander Frederick, reel 3.
101 TNA WO32/1302, Lees, Methods of prevention of venereal diseases, p. 1.
102 G. MacDonald Fraser, *Quartered Safe out Here: A Recollection of the War in Burma* (London: Harvill, 1992), pp. 136–7.
103 Richmond and Gear, 'The health of the Middle East force', p. 6.
104 Richmond and Gear, 'The health of the Middle East force', p. 6.
105 Imperial War Museums Art Collection, Art.IWM PST, 9063.
106 'Just another "Claptrap". You can ruin your future with – V.D!' Poster designed by Stacey Hopper to warn Allied troops in Italy about the dangers of venereal disease, 1943–44 ,Wellcome Archives, London, RAMC 466/48. See also Wellcome Images L0023970, http://wellcomeimages.org (accessed January 2012).
107 MacDonald Fraser, *Quartered Safe out Here*, p. 182.
108 IWM SA, 13230, Charles Bennett, reel 3.
109 Richmond and Gear, 'The health of the Middle East force', 5.
110 Harrison, *Medicine and Victory*, p. 153.
111 W. Franklin Mellor (ed.), *Medical History of the Second World War: Casualties and Medical Statistics* (London: HMSO, 1972), pp. 192, 282.
112 MacDonald Fraser, *Quartered Safe out Here*, pp. 182–3.
113 Wellcome Archives, GC/135/B.1/3, MacKeith, Some comments on the V.D. problem in an expeditionary force, p. 3.
114 IWM SA, 23367, Henry Butterworth, reel 2.
115 IWM SA, 24187, George Percy, reel 7.
116 IWM SA, 17630, Eric Murray, reel 4.
117 IWM SA, 30078, John Emerson, reel 3.
118 MacDonald Fraser, *Quartered Safe out Here*, p. 183.
119 S. Longden, *To the Victor the Spoils: Soldiers' Lives from D-Day to VE Day* (London: Robinson, 2007), p. 88.
120 TNA WO222/1302, Lees, Methods of prevention of venereal diseases, p. 2.
121 TNA WO222/1302, Lees, Methods of prevention of venereal diseases, p. 2.
122 Lt.-Col. A. Hood, 'The prevention of venereal disease with special reference to preventative ablution centres', *Journal of the Royal Army Medical Corps* 68:6 (1937), 390.
123 Hood, 'The prevention of venereal disease', 390.

124 War Office, *Manual of Military Law*, p. 24; TNA WO277/7, MacPherson, Discipline, appendix 1(a).
125 Court-martial records tell us only the numbers of men who were convicted and not those who were acquitted or who were suspected but never charged. It is also not possible to determine the exact number of men convicted of malingering, which was not defined as a crime under military law, but fell under the category of cowardice. TNA WO277/7, MacPherson, Discipline, appendix 1(a).
126 TNA WO277/7, MacPherson, Discipline, p. 51.
127 D. French, 'Tommy is no soldier: the morale of the Second British Army in Normandy, June–August 1944', *Journal of Strategic Studies* 19:4 (1996), 160.
128 TNA WO277/7, MacPherson, Discipline, pp. 50–1.
129 IWM SA, 12436, Ronald Sherlaw, reel 10.
130 R.M. Wingfield, *The Only Way Out: An Infantryman's Autobiography of the North-West Europe Campaign, August 1944–February 1945* (London: Hutchinson, 1955), pp. 114–16.
131 Wellcome Archives, RAMC/349, A 'malingerer's guide' to how to appear ill, disguised as a book of matches, one of the many distributed among British troops in Italy towards the end of the Second World War, 1944, pp. 11–12.
132 IWM SA, 12239, Gerald Barnett, reel 5.
133 French, *Raising Churchill's Army*, pp. 135, 138.
134 IWM SA, 13230, Charles Bennett, reel 3.
135 IWM SA, 14981, William Cornell, reel 8.
136 Longden, *To the Victor the Spoils*, p. 70.
137 IWM SA, 18255, Gordon Gent, reel 10.
138 IWM SA, 19867, John Buchanan, reel 7.
139 IWM SA, 19638, Charles Jordan, reel 10.
140 TNA WO71/1114, General Courts Martial, offence: desertion, charge sheet for Lieutenant-Colonel S. Cook, No. 1 Infantry Holding Battalion, 17 October 1945, p. 3.
141 TNA WO71/1114, Letter from Private Sidney Cook to Lieutenant Col. Collins, 1st Battalion, the Loyal Regiment, British Expeditionary Force France, 14 January 1940, sent from POW Camp in Brussels, Belgium, Prison a Saint Gilles, pp. 4, 7.
142 TNA WO71/1074, General Courts Martial, offence: desertion, statement by 2nd Lieutenant J. Speight, 4th Hussars, October 1942.
143 IWM SA, 22075, Kenneth Bond, reel 4.
144 IWM SA, 22075, Kenneth Bond, reel 4.
145 IWM SA, 23195, Roy Bolton, reel 6.
146 Richmond and Gear, 'The health of the Middle East force', 30.

147 M. Harrison, 'Medicine and the culture of command: the case of malaria control in the British Army during the two world wars', *Medical History* 40 (1996), 445–6.
148 IWM SA, 20149, Alex Gilchrist, reel 8.
149 IWM SA, 10956, Ernest Lanning, reel 4.
150 IWM SA, 30078, John Emerson, reel 3.
151 Wellcome Archives GC/135/B.1/3, Mackeith, Some comments on the V.D. problem in an expeditionary force, p. 4.
152 IWM SA, 17630, Eric Murray, reel 4.
153 IWM SA, 12239, Gerald Barnett, reel 4.
154 TNA WO277/7, MacPherson, Discipline, appendix 1(a).
155 French, 'Tommy is no soldier', 160.
156 IWM SA, 11207, James Bell, reel 3.
157 TNA WO277/7, MacPherson, Discipline, appendix 1(a).
158 TNA WO277/7, MacPherson, Discipline, p. 28.
159 D. French, 'Discipline and the death penalty in the British Army in the war against Germany during the Second World War', *Journal of Comparative History* 33:4 (1998), 538–43.
160 War Office, *Statistics of the Military Effort of British Empire during the Great War, 1914–1920* (London: HMSO, 1922), p. 648.
161 War Office, *Statistics of the Military Effort of British Empire during the Great War*, p. 667; TNA WO277/7, MacPherson, Discipline, appendix 1(a).
162 TNA WO277/7, MacPherson, Discipline, pp. 34–6.
163 TNA WO277/7, MacPherson, Discipline, p. 37.
164 French, 'Discipline and the death penalty'.
165 TNA WO32/15773, Auchinleck to War Office on the reintroduction of the death penalty, 7 April 1942, p. 1.
166 TNA WO32/15773, Telegram from Command Centre in the Middle East to War Office, 24 July 1942, p. 2.
167 French, 'Discipline and the death penalty', 541–2.
168 Harrison, *Medicine and Victory*, p. 2.

5

Fear, wounding and death

The overriding objective of the Allied forces in the Second World War was to fight the enemy: to defend areas under threat from Axis invasion and to liberate conquered territories.[1] The resources that the British Army used for this were essentially human. The front-line soldier, whose body had been honed and primed for combat, now took his chances in battle. There all his skills, training and experience would be put to the test, and there he faced the prospect of being wounded or killed. Neil McCallum was deployed with the Eighth Army in North Africa in 1942. He recalled in his memoir:

> It is the end of two years transformation from raw rookie in Britain to battle-reinforcement in Africa. This is the last metamorphosis and whatever emerges will be man, lunatic or corpse. One can sink no further into anonymity, be stripped not more of the idiosyncrasies of personality and taste. I am now what my civilisation has been striving to create for so long, a technically valuable, humanly worthless piece of flesh and blood, animate, responsive, and supposedly faithful until death.[2]

McCallum's testimony suggests that the body was expected to remain obedient and efficient even until death. Trained to respond automatically to orders and subsumed within the wider military machine, the body was the target of military discipline, even in the difficult circumstances of engagement with the enemy, where its damage and destruction were imminent.

However, even at the moment for which the soldier had long been prepared, the army was not content to leave his body alone. In a world characterised by lethal uncertainly, noise, danger and discomfort, men could easily succumb to a sense of fear that was intrinsically physical.[3] Joanna Bourke states that 'whatever a soldier's rank, fear was his persistent adversary. Its effects upon the body were particularly evident in wartime.' Fear manifested itself in trembling hands, sweating palms, sleeplessness, diarrhoea, chronic gastrointestinal problems, abdominal symptoms or the malfunctioning of the nervous system.[4] This chapter examines the methods used by the military authorities to counter these physiological effects of fear. It also outlines the army's medical administrative instructions and burial regulations that were designed to treat and dispose of the wounded and dead quickly and effectively. All of these procedures, I suggest, continued to control, organise and standardise men's bodies in the ever-present pursuit of military efficiency, even when they were injured or killed.

Yet, as we will see, in the difficult conditions of combat, the body could remain unruly and difficult to control. Faced with the breakdown or failure of their own bodies, or confronted with the sight of death in others around them, soldiers often struggled to regain control of their physical emotional selves. By exploring men's individual responses to fear, wounding and death, this chapter therefore uncovers an apparent paradox: the unpredictability of the body at the moment that it was meant to be most disciplined and ordered.

Finally, this chapter explores the pension provisions that were put in place during the war in order to compensate injured and dead bodies. While the care, treatment and rehabilitation of wounded and disabled servicemen have been examined elsewhere, the pensions system also highlights the continued intervention of the state in soldiers' bodies after demobilisation.[5] How awards were worked out, how soldiers applied for pensions and how they contested official decisions all demonstrate the state's surveillance of men's bodies after military service was over. These procedures show, furthermore, that the body remained a disputed terrain upon which struggles over control and resistance were fought out once soldiers had returned home.

Managing fear, wounding and death

While the impact of war on the body was perhaps most obvious through wounds or death, fear was also a major threat to effectiveness in combat, and its effects were undoubtedly physical. In 1946 Field-Marshal Viscount Montgomery recounted his wartime experiences in a speech on 'Morale in battle'. He stated that some soldiers 'under the stress of hardships or dangers surrender to fear'. This, he continued, resulted in 'timidity of action and slackness in appearance...those who have gone to seed will be dirty in their appearance and slovenly.'[6] The effects of fear could be more debilitating still. It caused sweating, shaking and the uncontrolled release of bodily fluids. While fear could be useful, something that men could convert into aggression and a spur to action, it could also have a paralysing effect, rendering soldiers incapable of performing their military roles.[7] D.K. Henderson, Professor of Psychiatry at the University of Edinburgh noted in 1941:

> Even amongst soldiers, a sort of man over whom all others it ought to have the least power, how often has it converted flocks of sheep into armed squadrons, reeds and bulrushes into pikes and lances, friends into enemies, the French White Cross into the Red Cross of Spain. Sometimes it adds wings to the heels, sometimes it nails them to the ground and fetters them from moving.[8]

Soldiers who were unable to overcome their fears were also at risk of developing more serious psychological conditions, such as hysteria, neurasthenia and anxiety states. The bodily manifestations of these were certainly profound. They included violent shaking, deafness, blindness, mutism and paraplegia.[9] Men suffering from these conditions constituted a serious threat to manpower. By 1941 around 1,300 psychiatric cases a month were being discharged from the army.[10] In theatres where soldiers were constantly bombarded by the weapons that they dreaded most or where they were separated from each other, between 10 and 20 per cent of all casualties were psychiatric cases.[11]

Aware of the serious risk that fear posed to the military mission, army leaders and medical staff used a range of measures to counter its harmful effects. As had been common practice in the First World War, some troops were issued with a rum ration before going into battle. This was more strictly controlled than it had been between 1914 and 1918 as it now required a medical officer's authorisation.[12] A quarter-pint measure, sometimes added to tea or cocoa, was believed to be enough loosen inhibitions but not to make men sick.[13] Army psychiatrist Lieutenant-Colonel MacKeith argued that it 'was highly likely to be good for a period of waiting and might be good for going over the top'.[14] Officer Henry Wilmott issued a 'tot' to his men before going into combat in Italy because he believed that it gave them 'Dutch courage'.[15] Another officer even referred to it as 'a battle winner'.[16] Without a doubt, soldiers who received a rum ration often experienced a beneficial effect. Private Ernest Harvey served in France and recalled that 'it put some heat into the body'.[17] Herbert Bedddows was issued rum onboard a landing craft

Fear, wounding and death

headed for the Normandy invasion in 1944. He recollected that 'it must have fortified me a little and that was very nice.'[18] Efforts to prevent more serious psychological disorders also often focused on the physical. Psychiatric centres were set up in forward areas at both divisional and corps levels. Here men who displayed psychoneurotic symptoms received a hot meal, as much hot sweet tea as they could drink, a bath, a change of clothes and, if needed, a sedative.[19] These measures clearly made a difference to the fighting spirit of the troops. Around 65 per cent of men treated at divisional centres were returned to full combatant duties in their original units within six or seven days.[20] At one centre on the outskirts of Kohima it was reported that most men 'improved out of all recognition with nothing more than sleep, food, a wash and a change of clothing'.[21]

Official strategies for managing the breakdown of the body were more formalised when it came to instances of wounding. During the course of the Second World War a total of 239,575 soldiers were wounded as a result of military service. This represented 86 per cent of all wounding

Figure 4 Men of the Norfolk Regiment receive their rum ration before going out on patrol in France in January 1940

cases in the three armed services.²² Wounding was not experienced equally among all troops and across all fronts. The war in Europe, where fighting was continuous, was the most costly in terms of battle casualties. Over 50 per cent of the men in five key regiments that served here were wounded in action.²³ Infantrymen were also the most likely to be wounded in war. In Normandy, for example, the infantry accounted for less than 25 per cent of the 21st Army Group, but suffered 71 per cent of its casualties.²⁴ If you were an officer, the outlook was bleaker still. In North-West Europe officers accounted for 26.5 casualties per 1,000 men every month, compared to 19.6 per 1,000 for other ranks. In the infantry almost three officers were wounded for every two ordinary soldiers.²⁵ Faced with such a loss of manpower, it was essential that the authorities retrieved, transported and treated the wounded as efficiently as possible so that men could be returned to the front, or at least be made fit enough to take on a non-combatant role along the lines of communication. Those soldiers who were no longer able to serve in any military capacity also had to be discharged as soon as possible back into civilian life, where they could make alternative contributions to the war effort.

The effective evacuation and treatment of casualties was achieved through a series of sophisticated medical posts that moved the body gradually away from fighting zones to supply bases. Within these worked teams of stretcher squads, field ambulances and motor ambulance convoys. First, a soldier would be taken to a regimental aid post (RAP), where his injuries were assessed and given immediate treatment. If necessary, he would be taken to an advanced dressing station, which functioned chiefly as a casualty collecting centre and only afforded treatment of the most urgent kind, such as the arrest of haemorrhage or the treatment of shock. From here, the soldier would proceed to a field dressing station, where he was inspected and given urgent treatment if necessary. He might then be taken to an advanced, or field, surgical unit, where he could be operated on. If his wounds were more serious the soldier would be sent to a casualty clearing station, where he would receive full medical or surgical treatment. At the earliest opportunity, he would then be sent to a general hospital, usually established at the base or along lines of communication and accessible to a port, for evacuation aboard carrier ship back to Britain.²⁶ Within this chain of evacuation there was a clear organisation of bodies, determined by the nature of the military environment and the severity of a man's wounds. In forward areas, where medical supplies and personnel could be scarce, it was particularly important that resources and efforts be directed to those patients who were most likely to recover. As such, medical personnel sorted and prioritised wounded bodies within a system known 'triage', or 'assessment according to quality'.[27] Under this procedure casualties were divided into three categories: those who were slightly injured and could be treated quickly and returned to duty; those who were wounded but likely to survive and who could be evacuated; and finally those who were considered too severely wounded to be transported. Men in this third group usually received the least medical attention and were often treated just so far as to ease their pain.[28] Joseph Day, who worked as a medical orderly in Egypt, recalled that staff did not treat 'lethal, mortal wounds, when men were beyond the possibility of medical treatment'. They would 'make them comfortable with morphia and sometimes gave lethal doses in order to stop the suffering'.[29]

Army doctors and surgeons frequently highlighted the significance of triage to the successful running of medical services in the field. In 1941 Colonel D. Stewart Middleton emphasised that 'the medical officer has an extremely important function to fulfil in classifying the wounded in order of priority for evacuation. It is useless to occupy transport by the dead and mortally wounded.'[30] Army surgeon John Watts, who worked in North Africa and France, also asserted in an interview that 'the most important thing is what's called is triage, which is sorting out those who must be dealt with in the forward area and those who are fit for forward evacuation, because you couldn't operate on everybody.'[31] In fact, the sorting of men's bodies began before they entered formal sites of medical treatment. Bert Wheeler, a stretcher bearer who served in Burma, had to choose which men to collect from the battlefield and take to the regimental aid post. He explained, 'you had to pick the ones you thought were the worst, who would live…say for instance he had shell or shrapnel in his chest, maybe the lungs exposed, then you wouldn't give him much hope. People with a broken leg, who can't walk, they had the…well more or less, first priority. People with other wounds, body-wise, which are not serious, they walked with you.'[32] The prioritisation of wounded bodies therefore took place at every stage in the chain of evacuation. From the battlefield to the hospital a clear hierarchy was put in place that ordered men's bodies according to their potential for restoration. This meant that all medical staff, from orderlies to surgeons, had to be highly selective in deciding which men should be treated.

Choosing which bodies to treat was not, however, always straightforward, even for the trained medical expert. Lieutenant Jean Limbosch worked in an advanced surgical unit in Belgium. He explained that 'firstly, the hopeless cases should be picked out. This is not as easy as one might think. Some hopeless cases, if seen early, look like having a chance and may not show any symptoms of severe shock until a number of hours after injury.'[33] Prioritising injuries could also be a difficult objective for the medical professional to reconcile. While working at a busy casualty clearing station in Normandy on D-Day, Watts was confronted with one particular case of abdominal wounding. This would take far longer to operate on than a more simple wound of the extremities and had only a 50 per cent chance of survival. Watts noted in his memoir, 'I had ten cases awaiting operation, all of them with a reasonable chance if they were operated on in time. This man's plight seemed so desperate that even if operated on he would have little chance of survival. Despondently I arranged for him to have a large dose of morphia to ease his pain, and instructed the stretcher bearers to put him in a corner to die. Then back to my cellar and the ten cases.'[34] In this instance, Watts faced a difficult organisational and ethical dilemma. As a doctor he felt a sense of obligation to the young solider, but as a military surgeon in a busy forward area, he had to concentrate his efforts on those bodies with greater potential for repair.

Clear optimums and limits were also set by the authorities for the numbers of men that could be efficiently treated in different environments and medical settings. In 1941 the government set up the Hartgill Committee to examine the army's medical services and provide suggestions for improvement. It recommended that each team working within advanced surgical centres should be able to deal with eight serious cases of wounding every eight hours, or rather work at a rate of one man per hour.[35] At a field dressing station in Normandy in 1944 Major F.S. Fiddes of the RAMC also observed that 250 cases was the maximum that could be handled within a twenty-four-hour period 'without loss of efficiency'.[36] In Normandy Watts noted that patients were processed at an optimum rate of '20–30 seconds per man'. So, when a man came in with an infected leg that he had been suffering from for two years, Watts explained that 'in order not to spoil our figures for the interval between wounding and admission we marked him as sick!'[37] Doctors were therefore clearly target driven and had bureaucratic systems to satisfy.

The organisation and processing of damaged bodies extended even into death. As was the case with wounding, the army suffered disproportionately compared to the navy and air force, with over half of all service deaths occurring among soldiers.[38] Moreover, advances in technology had introduced more destructive weapons, meaning that dead bodies were often not whole or intact. Whereas in the First World War 80 per cent of all wounds had been from gunshot, during the Second World War over 85 per cent of injuries were from mortars, grenades, aerial bombs, shells and mines. Unlike a clean bullet wound, these weapons often caused multiple injuries and removed whole areas of tissue and muscle.[39]

Describing the corpses on the battlefield after the fall of Tobruk in 1942, solider Harold Atkins recalled that they were 'deteriorating terribly, green, bloated, limbs hanging off, half off, no heads, half a body, goodness knows what'.[40] An effective system for disposing of dead bodies was therefore crucial, firstly, because these mutilated and decomposing bodies were a source of disease and, secondly, because the sight of the dead could undermine discipline and morale by confronting men with their own mortality.[41] Field-Marshal Montgomery noted that 'A corpse in a ditch or a grave by the side of the road will remind him [the soldier] of the peril of his position. He will suddenly realize that he himself is liable to be killed.'[42] The army pamphlet *Psychiatric Casualties: Hints to Medical Officers in the Middle East* also highlighted the negative impact that the military corpse could have on a soldier's state of mind. Among the 'special causes of stress' it listed 'the sight of the dead or of specially unpleasant wounds, all of these act as precipitating causal factors in both officers and men.'[43] Thus, only through its efficient burial and sanitisation could the authorities negate the corpse's ability to contaminate the bodies and the minds of other men in the field.[44] This work was organised by the Graves Service, which set all burial regulations, arranged for the provision of suitable cemeteries and ensured records were kept of all burials for identification purposes. In theatres abroad, graves registration units were also attached to each force and along lines of communication.[45] The task of clearing the dead and wounded was, however, down to the troops themselves, who had to dig graves and perform burials. In disposing of the dead, men had to follow a strict code of procedure, which was laid out in the Army's Field Service Regulations. First, all bodies were to be searched and, if possible, identified. The exact spot on which the corpse was found and the apparent date of death was then to be noted before each body was buried in a single, standardised grave of no more than 6 feet and 6 inches long, 2 feet wide and 5 feet deep, with no more than 1 foot between each one. Graves were to be marked with pegs and labels, or when not available, with bottles or tins, half-buried, open end backwards, with particulars of the burial written in black lead pencil on a sheet of paper or metal foil and placed inside.[46] It seems that in death, as in life, the soldier's body was to be homogenised, as each became subject to a standardised process of identification, documentation and disposal.

Experiences of fear, wounding and death

While the army devoted much attention to managing the damage and distress of the body in the combat zone, the testimonies of soldiers who experienced and witnessed fear, wounding and death show that they prompted a wide range of responses. In the face of action, some soldiers were able to repress the bodily manifestations of their fears as they came to focus their attention on the task at hand. George MacDonald Fraser, who first experienced armed combat while serving in Burma, noted in his memoir, 'at the moment of fixing bayonets I had that hollow feeling which most writers locate in the stomach but in my case manifests itself in the throat; after we were fired on I didn't notice it.'[47] As the authorities had recognised, fear was even useful, something that men could convert into fury and a spur to action. William Scroggie served as an officer in the Lovat Scouts in Italy. He first went into battle during the Allied advance on the Gothic Line in 1944. He recalled in an interview, 'I discovered that there is nothing more exhilarating in the world than combat. You behave as if your body belonged to someone else. You have no fear, no worries. You're on some kind of alternative high and it's just tremendously exciting and exhilarating.'[48] W.A. Elliot described a similar experience when he recounted an ambush in Italy:

> An awful savagery now seemed to take hold of us as we rushed along the embankment shouting oaths and shooting at Germans who were lying there. I felt as if some wild animal had got me by the throat and I had to keep shooting or else my normal self would return bringing fear along with it. There was even a savage pleasure in it. One German was truculent, refusing to double back down the line, and while we were arguing and threatening him, other Germans fired at us out of a trench. I shot him point blank; the effect was electric.[49]

Like William's, Elliot's story suggests an out-of-body experience. He felt as if a wild animal had got hold of him that kept his fear at bay. His fear was even transformed into pleasure as he continued with the attack, culminating in an 'electric effect' when he killed the enemy soldier. William Dunn, a tank driver hit by enemy fire on D-Day, also experienced his fear as something positive. He believed it restored function to his wounded body. Having been shot several times in the leg,

William suffered from five compound fractures. Yet he managed to run 50 yards to get out of the firing line. Explaining this in an interview, he stated, 'well, when you're frightened and there's bullets flying around, it makes you do queer things.'[50]

In the field of action, soldiers also developed their own bodily strategies in order to cope with their fears and anxieties. Tea, for instance, provided a crucial morale-booster. Described by one soldier as 'absolute, vital nectar', it replenished weary bodies, providing warmth, comfort and relief. Valued not only for its physical benefits, tea had other ready associations. 'Brewing up' became a central feature of life on the front line, taking soldiers to a good place as they were reminded of home and family.[51] D-Day veteran John Gray recalled that 'you'd die without a cup of tea, you know. You'd die fast without a cup of tea...I suppose it was a bit like home, if you like. It was a bit civil. Not civil in the sense of politeness but it was, it made you feel like a human being. It made you feel civilised to have a cup of tea.'[52] Soldiers also turned to smoking in order to calm their nerves. Cigarettes were issued to overseas troops in their ration packs and soon became much sought-after bartering items.[53] Describing the benefit that he derived from smoking, Bill Scott, an NCO in the Fife and Forfar Yeomanry, referred to it as 'a comfort, a solace... You smoked at every available opportunity and if your hand was shaking, that's all right.'[54] The experience of combat introduced many young men to their first cigarette and transformed them into long-term smokers. Bill Knights, who served with the Fife and Forfar Yeomanry in France, explained that he took up smoking because of 'the tension'.[55] Driver John Gray's tank was hit by enemy fire during an attack on the town of Cheux in Normandy in 1944. It was then that he smoked his first cigarette:

> We were both, Jimmy Byres and I, sitting there like a pair or jellies and I'm not kidding, we were shaking, the pair of us...He got this fag in his mouth. He lit it up, every bit of us was shaking. He got this fag in his mouth and in three puffs he was sitting there calm as anything. I said to him, 'Give us one of those, Jimmy,' and he gave me one, and I coughed everything out I could and I started to smoke, and virtually chain-smoked after that.[56]

Other soldiers were, however, unable to overcome the effects of fear on their bodies. Sent in as part of the Allied beach landings in Normandy on D-Day, William Spearman noticed how some men were unable to advance. He explained that 'you stay and die or you get off and live. People doing it for the first time, no matter how many times you tell them, they don't realise it, and nobody gets off the beach. Any one of us could tell you, they wouldn't get off. They were transfixed with fright. They couldn't get off. We were transfixed with fright but we had the knowledge that you either stopped and died or got off and got away.'[57] While William was able to convert his own fear into action in order to ensure his survival; he was unable to persuade the other men to do so. A similar story was told by James McCallum when he described the actions of one man in his unit during the Battle of the Mareth Line in Tunisia in 1943:

> One of the men in the company lay down and began to sob. He was one of the older men, almost forty, old enough to be the father of some of the youngsters with us. How to explain to him in the middle of battle that he should be up on his feet with his rifle in his hand – aggressively tough – and not spend his glorious patriotic hour lying on the ground moaning for his wife and family in the suburban home? There was no time to explain, to argue, to plead. He was beyond the direct order of oblique persuasion. He was left lying, the tears on his face mixed with sand and earth.[58]

In such instances the body was rendered useless, leaving men incapable of performing their military duties, even in the face of death. As McCallum noted, the man was supposed to be ready for action and 'aggressively tough', yet his emotion overrode his ability to continue with the task at hand. Recognising that he was beyond help, the other men in the unit had no choice but to leave him lying on the ground and continue without him. Even men who tried to repress the manifestations of their fears could be powerless to do so. Try as he might to stop, John Buchanan would 'involuntarily twitch with fear' before going into battle in France.[59] Patrol activity was a common feature of military service in all theatres in order to gain intelligence on the enemy. This was often a nerve-wracking experience for the men involved.[60] Describing a night patrol in Belgium, Rex Wingfield wrote in his memoir that 'we shivered. We shook. We shuddered. Our teeth chattered. We were bathed in sweat. Our muscles twitched and strained as we fought to stop ourselves from vomiting. It was no good. We leaned over the wall.'[61] Thus, it seems that these men's rational calculations were overwhelmed by their emotional bodies. At the moment for which their bodies had long been trained and prepared for, they became become most individualistic, betraying the instructions of military superiors and even the soldiers themselves.

This sense of detachment between body and self is also evident in recollections of wounding. Soldiers' accounts frequently describe a lack of physical pain at the moment of injury. Upon being hit by Japanese shellfire in northern India in 1944, Arthur McCrystal felt 'no pain as such…I mean, nothing would work but there was no knowledge of anything. I didn't know what had happened, in fact.'[62] Arthur Thompson, who was hit by enemy shelling in Normandy and lost his right leg, also asserted that 'When it's chopped off like that you don't feel any pain. I was amazed. No, in fact, as soon as I moved, that was when I knew I'd lost my leg because it dropped off, you see. You felt the bones grating, you see, but there was no pain with it. You felt the bone grating as it dropped off, you see, and it was just fastened on with skin at the back.'[63] Similarly, when describing being hit by Japanese shelling in Rangoon, W.A. Elliot noted that 'I found myself staring blankly at a small, smoking crater straight in front of my feet. Although my body must have hit the ground, I found myself in the curious position of thinking that I was looking down on myself. My reactions were "My God, I must be absolutely riddled", as I noticed in a detached way, blood beginning to flow down my left side.'[64] These narratives suggest that the men were unaware that they'd been wounded, or did not realise the extent of their wounds until they noticed clinical signs. Elliot only realised that he had been hit when he saw 'blood beginning to flow'. Arthur did not realise that he had lost his leg until it 'dropped off'. Although it is clear that injury brought their bodies into a heightened state of consciousness, there is also a sense that they became alienated from their own bodies, as their wounds were experienced as something 'other' to the self.[65]

Other veterans have focused on the events surrounding their injury rather than the injury itself. They have described with precision the times, places, dates and immediate events leading up to the moment of wounding. Typical of this is William Dilworth's testimony:

> Seven of us were detailed to go forward into no man's land and put a set of mines out between us and the German line. And we were coming up, and it must have been getting on for one o'clock in the morning, pitch black, and I'm walking along and I could just vaguely see the outline of the two wires that we'd already put, knowing that the minefield was further that side. And I'm walking backwards, making sure that this wire that we're pulling open doesn't catch against that that we've already laid, and then when it was fully open we'd all lift it up, you know, the other few would be along the length of it and lift it up on top of the other two so that there was a triangle of all these wires. And as I was walking back with the last coil of wire, thinking to myself we'll be going back to our line in a few minutes, out of the corner of my eye I saw two sparks, two red sparks. No noise, no nothing, just these two red sparks. The next thing I knew I was lying on the face of the earth.[66]

William was able to remember the task that he was assigned and the number of men who were with him on the night that he was injured. He described the movements that he made, the equipment that he used and the intended outcome of the event. In doing so he clearly set the scene of his wounding. According to Elaine Scarry, this sort of contextualisation is used in narratives of pain because of the difficulty of articulating pain itself. A description of an accident can convey the fact of the patient's agony more successfully than attempts to describe their pain directly.[67] Thus, by stressing the danger of the situation, that they were in 'no man's land' and it was 'pitch black', and by describing the sights and sounds of the weapon, that the 'sparks' came from the mine, William could convey the severity of his wounds. These temporal and spatial moments before his wounding appear to serve as reference points to describe the production of his injury, rather than the sensation of pain itself. When he regained consciousness, however, William became acutely aware of his body, as he started to search for the location and cause of his wounds. At this point his body was no longer in the background, but reappeared in the foreground of his experience.[68]

> I started putting my hands between my body; funny thing how you put your hands in to feel for your heart first. If you'd been hit in the heart you'd have been dead anyway so you wouldn't have been able to feel. It's a silly thing but it's automatic. So you feel, nothing there, pushing me hands in between the earth and me body, nothing wrong. I put my hands round the back and my hand sunk into my stomach, from the back. Well that's how it feels, because the nerve endings are so tender then that even a tiny hole feels as though it's a great big hole, you know. And I thought my hand had gone straight in and there was nothing, so I thought to myself, I'm gonna die. I've got no stomach or anything. I'm gonna die. I could imagine this great big hole in my back.[69]

William therefore came to experience his body in a disconnected or unfamiliar way. He searched for his wounds with his hands, suggesting that he could not sense any pain directly. Once found, he had a distorted view of his wounds, that there 'was a great big hole' in his back. His locating of his wounds brought his body into a heightened state of consciousness as he came to anticipate the extent of the damage caused.

It was not just the event of wounding but how a man coped with his injuries that created distance between body and self. To cry was both unsoldierly and unmasculine, and men often struggled to prevent such an outward flow of emotion. Leslie Perry lost his left leg and his right arm after being blown up by a mine in Normandy. When he

Fear, wounding and death

was collected by the stretcher bearers he began to sing because 'I was frightened to make myself look stupid by crying. I didn't want to cry. I felt like I wanted to but I didn't want to. I didn't want to let myself down.'[70] Leslie seems to have worked hard to maintain control over his body in order to maintain the appearance of stoic masculinity.[71] Officer Ernest Lanning, on the other hand, did cry after being shot in the leg while serving in North Africa in 1942. He recalled, 'my mind was perfectly clear except, "what an undisciplined, ridiculous thing to do, stop." But I couldn't stop it.'[72] Clearly, Ernest experienced a sense of conflict between his body and mind. Although he desperately wanted to stop this embarrassing, and 'undisciplined' performance, and willed himself to do so, in the end he could not take command of his body in order to stop crying.

Despite the army's efforts to regulate and sanitise death, individual encounters with corpses were also still especially difficult. Some men were taken aback when confronted with the destruction caused to bodies by war. Ronald Petts, an NCO with the 224th Parachute Field Ambulance, served with the RAMC in North-West Europe and Palestine. He described the death of a young soldier hit at close range by a grenade. He stated, 'I was completely shocked. I opened up his tunic and his gut was completely smashed. As I opened up his tunic and belt and trousers his whole gut just fell away.'[73] Central to his recollection was the collapse between the internal and external body. Perhaps this was simply because of its gruesomeness, or maybe because it contradicted so greatly with the ideal 'whole' body of the heroic soldier.[74] Confrontations with mutilated and dismembered bodies could have a deeper impact still. Chris Shilling suggests that death radically alters the accepting attitude which is normally adopted in everyday life. It challenges people's sense of what is real and meaningful about their embodied selves and the world around them.[75] This was the case for William Dilworth, a member of the Salvation Army, who lost his religious beliefs after being sent to clear the dead from the battlefield at Anzio in 1944:

> There was one leg...British Army. There was one leg with the boot and the gaiters and everything, you know, sticking out from under this bush. So I bent down, got hold of the ankle and tugged to pull the whole body out from under the bush where it had been blown, and I just fell over backwards holding from the knee down in my hand. And I looked at the stump where it was all raw and bloody and everything and I looked up and I said, "There's no", and I swore, "God," and from that, from that moment, I've never been religious. And I packed up religion altogether then, because

> I thought to myself, to allow this to happen to anybody, don't matter even if its enemy or not, there can't be a God, so from now on I'm not a religious man.[76]

William's reaction was prompted by the severe damage that the body had endured. It seems that he was prepared for the task of clearing dead bodies but expected them to be intact or complete. The sight of a 'stump' that was 'all raw and bloody' so shocked him that he came to question the cause for which he was fighting and ultimately lost his religious convictions.

As the authorities had anticipated, the corpse also proved to be a visible reminder for the soldier of his own embodied existence, or rather his own potential death.[77] For D-Day veteran Leonard Harkins the sight of the dead on the beaches came as a real shock, 'especially when you look in a tank and see bodies like that and thinking, well, tomorrow you could be the same'.[78] Assigned to collect the dead after battle in North Africa, Harold Atkins explained that 'this was possibly the one of the most unpleasant tasks that I or I think anybody has to perform, particularly if you are an infantryman and you're still going to do some more fighting…you're aware that there by the grace of God go I'.[79] Rex Wingfield described a similar response when his unit encountered two corpses after a patrol in Belgium. The second, the body of a corporal, had the most profound effect:

> We found the body in a ditch where it had been blown by the blast. The hand still clutched a rifle. This was the first body of ours that we had seen. We hesitated, half afraid, half curious. The patrol survivor quickly and reverently placed the body on the tank, hands dangling over the side. We found the Corporal three hundred yards further on, slumped on the bank where the world had erupted in his face. The first corpse shook us by its naturalness when we expected mangled, bloody wounds, but here was the dreadfully smashed body all of us saw as ourselves in our nightmares. Usually nothing can be seen, as the uniform covers all wounds except those on the hands and face. This was the face.[80]

Again it was the visibility of the wounds that shocked the men and reminded them of their own mortality. While they were surprised by the 'naturalness' of the first corpse, which had suffered little damage, the facial injuries of the second confronted them with their own potential bloody deaths. It seems that the men did expect dead bodies to be mutilated, to encounter 'mangled bloody wounds', but despite this they were still deeply affected by the sight of the dead corporal.

Fear, wounding and death

As service life continued, however, and death was an everyday reality, some men did become accustomed to seeing dead bodies. Private Rubin Wharmby, who served with the 1st Battalion Royal Lancashire Regiment in North-West Europe, recalled that 'you get used to it. At first you see them and, you know, it's sick, and then you don't bother.'[81] For William Tichard, a gunner in the 83rd Field Regiment, Royal Artillery in Normandy, the shock of seeing dead bodies soon wore off as he focused all his attention on the task at hand. He explained that 'you were getting used to seeing that, dead people. You know, it grew on you...It was heartbreaking but you had a job to do.'[82] This could become a useful coping mechanism for men confronted regularly with death. George MacDonald Fraser described the reaction of his unit to the deaths of two fellow servicemen while serving in Burma:

> There was no outward show of sorrow, no reminiscences or eulogies, no Hollywood heart-searchings or phoney philosophy...It was not callousness or indifference or lack of feeling for two comrades who had been alive that morning and were now names for the war memorial: it was just that there was nothing to be said. It was part of war; men died, more would die, that was the past and what mattered now was the business in hand; those who lived would get on with it. Whatever sorrow was felt, there was no point in brooding about it, much less in making, for form's sake, a parade of it. Better and healthier to forget about it, and look to tomorrow.[83]

Like Tichard, George and his men had come to concentrate first and foremost on the military mission. While they were upset by the deaths of their friends, they had learned not to dwell on it but rather to focus on the task to which they had been assigned. The men had not necessarily become desensitised to death but had developed a way of dealing with it.

Soldiers also found practical uses for the bodies of the dead. Rubin Wharmy claimed, 'we used to use them for landmarks.'[84] The corpses that William Corbould encountered on his way to Monte Cassino also constituted makeshift signposts. He explained that 'We were told, "Turn right at the dead New Zealander. Keep going until you come across an Italian. Turn right at the next German," et cetera, et cetera, et cetera. It sounds as if I'm being very crude. It sounds as if I'm not telling the truth, but that is in fact how we found our way.'[85] In the African desert, with only rocks to sit on, the men in Neil McCallum's unit used dead bodies as chairs and even worked out the most comfortable way of sitting on them. McCallum noted that 'to use a corpse as a seat it should be turned on its face. It is difficult to sit on a dead chest, and besides, the face is obtrusive. But the small of the back is rather like a saddle.'[86] It seems to have been the faces of the bodies that made them human. Once turned around, they were converted into furniture of the combat theatre, simply part of the scenery of war.

Commodifying wounded and dead bodies

Official interest in dead and wounded bodies did not end on the battlefield or in the army hospital. Through its military service the injured body and the corpse also became a commodity to be paid for by the state.[87] At the outbreak of the war, the Pensions (Navy, Army, Air Force and Mercantile Marine) Act was passed, which transferred the powers and duties of the service departments to the Ministry of Pensions.[88] By the end of the conflict, over 300,000 male soldiers had received first awards.[89] The system by which men were assessed for disability pensions dated back to 1917 and was based on the principle that an individual's disability had to be 'attributable to military service during the war'.[90] The rates awarded were assessed purely according to physical criteria, with no regard for the man's civilian occupation or personal circumstances:

> The degree of disablement of a member of the Military Forces shall be the measure of disablement (expressed by way of a percentage, one hundred degrees representing total disablement) which is certified to be suffered by that member by a comparison of his condition of a normal healthy person of the same age and sex, without taking into account the earning capacity in his disabled condition of that member in his own or any other occupation, and without taking into account the effect of any individual factors or extraneous circumstances.[91]

Thus, disability was based purely upon loss of faculty. A soldier who had lost both hands, for instance, was categorised as 100 per cent disabled, while the loss of vision in one eye or the amputation of a leg below the knee was measured at 40 per cent disability.[92] While all servicemen were eligible to apply for pensions, the rates payable varied considerably according to military rank.[93] Tables 3 and 4 below show the rates for non-regular soldiers and officers in September 1939.

These figures show that at every level of disability a clear hierarchy was in place. The body of a colonel was valued at twice as much of that as a captain and almost four times as much as a Class V, or lowest ranked, private soldier. Taking for example, men assessed at the highest level of disablement. A Class V soldier was entitled to 32 shillings and 8 pence per week, which equated to approximately £84 per year. A captain would

Fear, wounding and death

Table 3 Weekly rates of disablement pension (s.d.) for non-regular soldiers, by rank and degree of disablement, September 1939

Percentage degree of disablement	Warrant officer Class I	Class I	Class II	Class III	Class IV	Class V
100	45	42.6	40	37.6	35	32.6
90–9	40.6	38.3	36	33.9	31.6	29.3
80–9	36	34	32	30	28	26
70–9	31.6	29.9	28	26.3	24.6	22.9
60–9	27	25.6	24	22.6	21	19.6
50–9	22	21.3	20	18.9	17.6	16.3
40–9	18	17	16	15	14	13
30–9	13.6	12.9	12	11.3	10.6	9.9
20–9	9	8.6	8	7.6	7	6.6

Source: Taken from *Royal Warrant for the Retired Pay and Pensions etc. of Members of the Military Forces Disabled, and of the Widows, Children and Dependants of Such Members Deceased in Consequence of the Present War* (Cmd. 6105), p. 10.

Table 4 Yearly rates of disablement pension (£) for non-regular officers, by rank and degree of disablement, September 1939

Percentage degree of disablement	Colonel or higher	Lieutenant-colonel	Major	Captain or subaltern
100	300	250	200	150
90–9	270	225	180	135
80–9	240	200	160	120
70–9	210	175	140	105
60–9	180	150	120	90
50–9	150	125	100	75
40–9	120	100	80	60
30–9	90	75	60	45
20–9	60	50	40	30

Source: Taken from *Royal Warrant* (Cmd. 6105), p. 15.

be awarded an annual sum of £150, while a colonel would receive £300.[94] These figures progressively increased throughout the war, always with the same level of distinction between ranks. By the end of the conflict, the rate of pension for a Class V private assessed at 100 per cent disability was roughly £104 per year, while a subaltern was entitled to £210 and a colonel or higher £420.[95]

Set values were also attached to specific bodily parts. The amounts payable again varied by rank but with fewer graduations. Minor disabilities were categorised simply into two groups, 'officer' and 'other rank'. For example, the loss of the whole right finger would award an officer £120 while an ordinary soldier received £60. For the loss of a big toe an officer was awarded £80 and an ordinary soldier £40.[96] Again, therefore, the body of an officer, or rather his respective bodily parts, were valued at exactly twice as much as those of a man in the ordinary rank and file. These sums were paid irrespective of the branch or arm of the service in which a soldier had served. Thus, an infantryman would receive the same compensation for the same level of disability as an artilleryman, as long as both held the same army rank.

Following this trend, dead bodies also acquired a financial value that maintained distinctions between the ranks. Under wartime provisions, the wives or dependants of army personnel who had died as a result of service could claim a military pension. The wife of a deceased private soldier was entitled to 35 shillings per week, or £91 per year, if she was less than forty years of age or had a dependent child. The wife of a captain, in the same circumstances, was entitled to £150 per year. A deceased general's wife, however, was entitled to £540 per year, irrespective of personal circumstance. Officers' wives were also eligible to receive a lump sum death gratuity, applicable if their husbands had been killed in action, on flying duty, or had died from war-related wounds within seven years.[97] These one-off payments ranged from £300 for a major's wife to £1,500 for that of a general.[98] Once more, military hierarchy was the main organisational device in the ways that men's bodies were commodified during the war.

State provision for dead and wounded bodies was not, however, unique to the wartime military. By the close of hostilities, over 17,000 civilians were also receiving pensions for injuries or deaths caused by enemy attacks at home.[99] In September 1939 the government introduced the Personal Injuries (Civilians) Scheme. Administered by the Ministry of Pensions, the scheme applied to the employed, the unemployed, children, housewives and pensioners.[100] As was the case for the armed services, disability was based purely on loss of faculty, which was calculated into a percentage. A government pamphlet stated in 1944 that 'The degree of disablement is assessed by the Ministry by making a comparison between the condition of the injured person as disabled by the injury and the condition of a normal healthy person of the same age and sex, without taking into account the earning capacity of the disabled person in his own or any other occupation.'[101] During the war the rates awarded to injured civilians were consistent with those for the lowest-ranked soldiers. In October 1939, for example, the weekly rate for a man who was suffering from 100 per cent disablement was 32 shillings and 6 pence, the same as for a Class V private.[102] By 1944 this rate had increased to 40 shillings per week, again for soldier and civilian alike.[103] Compensation in cases of death was also calculated along military lines. In 1944 the weekly rate for a widow over forty years old, or with eligible children, or who was incapable of self-support, was 32 shillings and 6 pence. For all other cases the weekly rate was 20 shillings.[104] These were the same as the rates payable to the wives and dependants of Class V soldiers in the army.[105]

The values attached to military and industrially disabled bodies also came to align during the war. Since 1897 workmen injured in the course of their employment had been compensated through the Workmen's Compensation Scheme.[106] This system had originally applied only to dangerous employments, such as railways, factories, mines, quarries and engineering works. An amendment Act of 1906 had, however, extended the scheme to cover all persons working under a contract of service. It provided for death or disablement caused by scheduled industrial diseases.[107] The rates payable had been based on the individual's earnings. Before the war the maximum for incapacity was 50 per cent of an individual's pre-accident weekly earnings.[108] Government legislation introduced during the conflict raised compensation ceilings to two-thirds of the worker's average weekly earnings for cases of total incapacity, thus bringing the rates payable closer to military provision. By 1943, for instance, a single man injured at work could receive up to 35 shillings a week for the first 13 weeks and 40 shillings thereafter.[109] These were the same rates of disability pension being paid to a Class V soldier suffering from 100 per cent disability.[110]

Indeed, it was during the Second World War that demands arose within the government to reform workers' compensation so as to create a system closer to that of war pensions. In June 1943 a Workmen's Compensation Advisory Committee was established under the leadership of Home Secretary Ernest Bevin. Influenced by the Beveridge

Report of 1942, which emphasised insurance and non-means-tested subsistence benefits, it recommended uniform flat rates, paid for by contributions from employer, worker and the state.[111] Upon injury an allowance would be paid to workmen, followed by an industrial pension. The amount awarded was based on the extent of disablement, measured 'by comparison with a normal healthy person of the same age and sex'.[112] This echoed the principle applied to wounded military personnel. In fact, the desire to treat workers and soldiers as one was made explicit:

> This system is in many respects like that which is the basis of war pensions schemes. It thus recognises a certain similarity between the position of the soldier wounded in battle and that of the man injured in the course of his productive work for the community. Neither is liable to have his pension reduced on account of what he may earn after the injury; each is compensated not for loss of earning capacity but for whatever he has lost in health, strength and the power to enjoy life.[113]

The Committee proposed an injury allowance, payable for thirty weeks, of 35 shillings per week for a single man. This was followed by an industrial pension that varied according to degree of disablement, again ranging from 20 to 100 per cent. The rates payable were then equivalent to the pension of a disabled Class V soldier.[114] In cases of death the worker's wife or dependants would be awarded weekly pensions, also at the same rates as those paid to the lowest-ranked soldiers' widows.[115] These reforms were eventually achieved through the National Insurance (Industrial Injuries) Act of 1946.[116]

Yet, while the government did start to dismantle the boundaries between the civilian and the military spheres during the war, in many respects the soldier's body continued to be considered as distinct from that of the worker.[117] Firstly, the flat-rate benefits paid to all civilians irrespective of earnings or social class reaffirms the importance of rank as a significant organising tool in the army. While wounded civilian bodies were standardised, military bodies continued to be ordered within a formal hierarchy that did not exist in civil society. Secondly, the most that injured civilians could ever hope to achieve were the same rates as the lowest-ranked soldiers. The variations in pension rates for servicemen allowed all men above Class V private to be awarded more than injured citizens or industrial workers. As such, many soldiers' bodies were prized over their civilian counterparts. Indeed, Peter Bartrip has argued that the benefits paid under the Personal Injuries (Civilians) Scheme were calculated according to a notion that civilians should fare no better than servicemen, who were admitted only after medical examination and were subject to military discipline for the whole length of service. This control justified the provision of disability pensions to servicemen at a rate of benefit superior to that applicable to the rest of the workforce.[118] It therefore appears that the wounded army body was not only valued in terms of malfunction, but also as a healthy entity, for the role it had performed and the discipline endured before incapacity.

Nevertheless, the system of compensating wounded soldiers was not without complication. As had been revealed in debates over physical selection for entry into the army, the disability pensions issue became the focus of much controversy as doctors, politicians and military personnel debated over which bodies should be eligible for awards. From the outset of hostilities, men applying for pensions had to prove that that their disability was directly attributable or materially aggravated by war service. This required a high standard of medical evidence, and a large number of cases were refused in the early years.[119] As the war continued this became a major point of dispute among politicians. Several cases were raised in parliament of men who had been classed as A1 upon recruitment but who, after being invalided out, were refused a pension on the grounds that their disability had not been contracted during service.[120] In May 1942, for example, MP Rhys Davies challenged the government's policy of granting pensions based on medical evidence alone. He noted that in his constituency of Westhoughton, 16,000 young men had been recruited into the forces from an approved society. Eight hundred of these men were later discharged on medical grounds, 142 of whom had been passed as A1 upon recruitment but had not been granted pensions. This was despite the fact that according to the society's records they had never been unhealthy until they were enlisted into the forces.[121] Similar stories were reported in the press, such as the case of J.C. Foate, which was published in the *Daily Mirror* in January 1941. After twenty years of army service, he was discharged as unfit but did not qualify for a disability award because, according to the Ministry of Pensions, his disability had been sustained while he was in the army before the outbreak of hostilities. This was despite the fact that he had been passed as fit for active service during the current war. The newspaper concluded that 'the Treasury has a cash register where a heart should be.'[122]

Amidst such bad publicity, in 1943 the Ministry of Pensions made several alterations to pension provision which transferred the onus of proof from the individual to the state. Rather than the claimant having to prove that his condition was the result of, or aggravated by, war service, the Ministry now had to prove that this was not the case. The government conceded that a man's medical grading on entry into the armed forces should be taken as evidence that 'he was fit for the kind of service demanded of him in that medical category.' As such, it concluded that, 'in the event of his being subsequently discharged on medical grounds any deterioration in his health which has taken place is due to his service.'[123]

Disputes between individual soldiers and the government representatives nevertheless continued and were played out in appeals tribunals. These had been a feature of the First World War and reappeared in 1943 as a result of the new government legislation. Designed to deal with both entitlement and assessment issues, each tribunal consisted of a legal member as chairman, a medical member and a lay member. Each presentation to a tribunal required the preparation of a reasoned 'Statement of Case' setting out all the relevant information, including extracts from service documents dealing with medical and other history, the appellant's own supporting statement and the Ministry of Pension's reasons for not admitting the claim. Both pension applications and appeals could be registered at any time, meaning that the body could be a site of contestation long after military service had ended. Tank Corps veteran Roy Tomalin suffered from loss of hearing and applied for a pension in the 1990s. He recalled that 'It went on for years and they wouldn't agree. And in the end they got back and they said no. They couldn't agree that my service life had anything to do with my loss of hearing. They would give me a hearing aid and that would be it.'[124] The pensions system also meant that the state's monitoring and surveillance of men's bodies continued after they had returned to civilian life. Ministry of Labour medical boards regularly inspected those in receipt of an award. Roy did get a small pension for a chest wound acquired in France, but later had this stopped. Recalling his last medical examination, he stated, 'I had to go to Chelmsford and they gave me a medical. They stuck a few pins in me and because I didn't ouch loud enough they said it had healed itself.'[125] Norman Marshall had his left arm amputated during the war and was awarded a disability pension straight away. Yet for years after the war he still had to travel to Liverpool for annual medical examinations. According to Norman, 'their idea seemed to be they wanted to see if I'd grown another arm, I reckon, because all you seemed to do was to go in and they'd look at you and say, "Oh, yes", you see, "Off you go."'[126] Stories like these tell us that while disabled soldiers' bodies were often privileged over their civilian counterparts through more generous pensions provisions, compensation did not always come easily. As a result of its military service, the body continued

Fear, wounding and death

be scrutinised, measured and assessed by a suspicious state. The body also remained a site of contest and negotiation between soldier and state long after the war was over.

Conclusion

The chaos and the destruction of the body were of crucial concern to the British military authorities during the Second World War as part of the enduring quest for manpower. Aware that fear could seriously undermine the fighting capabilities of individual men, officers and medical staff issued rum, provided hot meals, clean clothes and opportunities to rest. They did not simply hope that men would resist their fears, but tried bolster and reinforce their bodies against its harmful effects. When bodies were wounded or killed, they were subject to more formal practices of control and regulation in similar ways to those imposed on able bodies in the military during the war. Wounded soldiers entered an official sorting process in which their bodies were categorised and treated according to their military usefulness. In the field, dead bodies were searched and identified according to a set code of instructions. Even the sizes of graves made no effort to memorialise the individual. In these moments of problematic performance, or indeed even when it had expired, the authorities did not leave the body alone, but continued to impose a series of regulatory controls upon it.

Combat was also, however, the moment when the army's efforts to order and discipline the body were tested against the impulses of individual men. Soldiers' accounts of battle reveal that fear, wounding and death often made bodies impossible to control. For some the fear of death or the sight of the dead and wounded was paralysing, while others shivered, shook and vomited. Despite their training, many of these men were ultimately unable to perform the duties for which they had been prepared. Some men wanted to repress the physical manifestations of their anxieties, but found that they were powerless to do so. Men who were wounded also experienced their bodies as something 'other'.[127] The sense of no physical pain or being detached from one's own body is common in stories of woundings. Some men did not even realise that they had lost limbs until they saw their wounds. The sight of dead and incapacitated bodies also confronted men with their own fragile existence. The sight of bits of bodies and flesh challenged their perceptions of the war around them. As such, the body was not just a predictable and efficient machine that could be controlled by the army or the individual soldier. Bodies could fail. They could prevent action and even disrupt fixed notions of the self.

Beyond the battlefield the state continued to de-individualise wounded and dead bodies through a complex series of disability pension rates. Although this system provided a wide range of provision for various types of injuries, it preserved significant distinctions between officers and ordinary soldiers in rates of compensation awarded. This even applied to the body's respective components, as the finger or the toe of an officer was worth exactly twice as much as that of an ordinary soldier. The compensation issue also linked the military body once more to its civilian counterpart, as the rates payable for servicemen, citizens and industrial workers came to align more closely during the war. This, however, was by no means an even system, as the state maintained a considerable gap between all soldiers above the very lowest rank and the civilian disabled. This shows that many bodies continued to be privileged as a result of their military participation. Yet, the pensions system was not always straightforward and easy for men to negotiate. Debates in parliament and veterans' testimonies reveal that many did not get the compensation they felt they were entitled to. Some men fought for years at tribunals. Even those who were given awards faced routine inspections by Ministry of Pensions doctors to make sure that they were still disabled. While men were therefore entitled to support from the state by virtue of their experience as soldiers, they were not uniformly successful in their appeals. For many, their bodies continued to be treated with suspicion long after their military service was over.

Notes

1 Ellis, *The Sharp End*, p. 57.
2 McCallum, *Journey with a Pistol*, p. 45.
3 French, *Raising Churchill's Army*, p. 135.
4 J. Bourke, *Fear: A Cultural History* (London: Virago, 2005), pp. 199–200.
5 Anderson, *War, Disability and Rehabilitation in Britain*; D. Gerber, 'Disabled veterans, the state, and the experience of disability in Western societies, 1914–1950', *Journal of Social History* 36:4, (2003), 899–916.
6 Field-Marshal Viscount Montgomery, 'Morale in battle: address given to the Royal Society of Medicine', *British Medical Journal* (9 November 1946), 703.
7 Bourke, *Fear: A Cultural History*, pp. 199–200.
8 D.K. Henderson, 'The significance of fear', *Edinburgh Postgraduate Lectures in Medicine* 2 (1941), 29.

Fear, wounding and death

9 General Headquarters, Middle East Forces, *Psychiatric Casualties: Hints to Medical Officers in the Middle East Forces* (Cairo: Middle East Forces, 1942), p. 5.
10 B. Shephard, *A War of Nerves: Soldiers and Psychiatrists, 1914–1994* (London: Pimlico, 2002), p. 188.
11 Crew, *The Army Medical Services: Administration, Volume I*, p. 489.
12 E. Jones and N.T. Fear, 'Alcohol use and misuse within the military: a review', *International Review of Psychiatry* 23 (2011), 167–8.
13 Longden, *To the Victor the Spoils*, pp. 68–70.
14 S.A. MacKeith, 'Presentation to Northern Command, York, 3 March 1945', unpublished report cited in Jones and Fear, 'Alcohol use and misuse in the military', 168.
15 IWM SA, 22666, Henry Wilmott, reel 3.
16 IWM SA, 24921, John Coldwell-Horstall, reel 5.
17 IWM SA, 14977, Ernest Harvey, reel 7.
18 IWM SA, 20372, Herbert Beddows, reel 2.
19 TNA WO32/11550, Report on a conference on psychiatry in forward areas, p. 4.
20 Crew, *The Army Medical Services: Administration, Volume I*, p. 489.
21 TNA WO32/11550, Report on a conference on psychiatry in forward areas, August 1944, p. 5.
22 *Strength and Casualties of the Auxiliary Services of the United Kingdom* (Cmd. 6832), p .8.
23 Ellis, *The Sharp End*, p. 162.
24 French, *Raising Churchill's Army*, p. 147.
25 French, *Raising Churchill's Army*, pp. 77, 148.
26 Crew, *The Army Medical Services: Administration, Volume II*, pp. 458–68.
27 Lt.-Col. Debenham and Lt.-Col. A.B. Kerr, 'Triage of battle casualties', *Journal of the Royal Army Medical Corps* 84:5 (1945), 125.
28 J.E. McCallum, *Military Medicine from Ancient Times to the 21st Century* (Santa Barbara: ABC-Clio, 2008), p. 330.
29 IWM SA, 12412, Joseph Day, reel 5.
30 Col. D. Stewart Middleton, 'The work of a regimental medical officer', *Journal of the Royal Army Medical Corps* 16:6 (1941), 316.
31 IWM SA, 18364, John Watts, reel 1.
32 IWM SA, 20897, Bert Wheeler, reel 5.
33 Lt. J. Limbosch, 'Problems of surgery in the field', *Journal of the Royal Army Medical Corps* 82:5 (1944), 212.
34 J.C. Watts, *Surgeon at War* (London: Allen and Unwin, 1955), p. 93.
35 Crew, *The Army Medical Services: Administration, Volume II*, p. 473.
36 Maj. F.S. Fiddes, 'The work of a field ambulance in the Battle of Normandy', *British Medical Journal* (31 March 1945), 448.
37 Watts, *Surgeon at War*, p. 93.

38 In total there were 264,443 deaths in the British armed services during the Second World War. Of these, 144,079 occurred in the army. *Strength and Casualties of the Auxiliary Services of the United Kingdom* (Cmd. 6832), p. 8.
39 Ellis, *The Sharp End*, p. 178.
40 IWM SA, 12440, Harold Atkins, reel 7.
41 E. Hallam, J. Hockey and G. Howarth, *Beyond the Body: Death and Social Identity* (London: Routledge, 1999), p. 128.
42 Montgomery, 'Morale in battle', 703.
43 General Headquarters, Middle East Forces, *Psychiatric Casualties: Hints to Medical Officers*, pp. 6–7.
44 See Shilling, *The Body and Social Theory*, p. 153
45 War Office, *Field Service Regulations, Volume I: Organization and Administration* (London: HMSO, 1930), pp. 114–15. These orders were still in operation during the Second World War.
46 War Office, *Field Service Regulations, Volume I*, p. 376.
47 MacDonald Fraser, *Quartered Safe Out Here*, p. 86.
48 IWM SA, 13281, William Scroggie, reel 3.
49 Elliot, *Esprit de Corps*, p. 35.
50 IWM SA, 12938, William Dunn, reel 2.
51 IWM SA, 19802, Bill Scott, reel 3.
52 IWM SA, 20202, John Gray, reel 7.
53 TNA WO203/812, Cigarettes: supply and complaints, 1944.
54 IWM SA, 19802, Bill Scott, reel 3.
55 IWM SA, 21015, Bill Knights, reel 2.
56 IWM SA, 20202, John Gray, reel 8.
57 IWM SA, 9796, William Spearman, reel 4.
58 McCallum, *Journey with a Pistol*, p. 89.
59 IWM SA, 19867, John Buchanan, reel 6.
60 Elliot, *The Sharp End*, p. 53
61 Wingfield, *The Only Way Out*, p .45.
62 IWM SA, 19066, Arthur McCrystal, reel 3.
63 IWM SA, 13370, Arthur Thompson, reel 1.
64 Elliot, *Esprit de Corps,* p. 117.
65 G. Bendelow and S. Williams, 'Pain and the mind–body dualism: a sociological approach', *Body and Society* 1:2 (1995), 83–103.
66 IWM SA, 18435, William Dilworth, reel 5.
67 E. Scarry, *The Body in Pain: The Making and Unmaking of the World* (Oxford: Oxford University Press, 1985), pp. 14–15.
68 Shilling, *The Body and Social Theory*, p. 184.
69 IWM SA, 18435, William Dilworth, reel 5.
70 IWM SA, 20009, Leslie Perry, reel 4.
71 J. Hopton, 'The state and military masculinity', in Higate (ed.), *Military Masculinities*, p. 112.

72 IWM SA, 19056, Ernest Lanning, reel 6.
73 IWM SA, 9732, Ronald Petts, reel 16.
74 Jarvis, *The Male Body at War*, p. 89.
75 Jarvis, *The Male Body at War*, p. 155.
76 IWM SA, 18435, William Dilworth, reel 5.
77 Hallam, Hockey and Howarth, *Beyond the Body*, p. 133.
78 IWM SA, 21109, Leonard Harkins, reel 4.
79 IWM SA, 12440, Harold Atkins, reel 7.
80 Wingfield, *The Only Way Out*, p. 84.
81 IWM SA, 18741, Rubin Wharmby, reel 3.
82 IWM SA, 20620, William Tichard, reel 3.
83 MacDonald Fraser, *Quartered Safe out Here*, pp. 88–9.
84 IWM SA, 18741, Rubin Wharmby, reel 3.
85 IWM SA, 23216, William Corbould, reel 4.
86 McCallum, *Journey with a Pistol*, p. 107.
87 Bourke has also shown this in the case of the First World War. Bourke, *Dismembering the Male*, p. 43.
88 A. Sandison, 'The medical services of the Ministry of Pensions', in MacNalty and Mellor (eds.), *The Civilian Health and Medical Services: Volume II*, p. 152.
89 Sandison, 'The medical services of the Ministry of Pensions', pp. 216–17.
90 *Royal Warrant for the Retired Pay and Pensions etc. of Members of the Military Forces Disabled, and of the Widows, Children and Dependants of Such Members Deceased in Consequence of the Present War* (Cmd. 6105), p. 8.
91 *Royal Warrant for the Retired Pay and Pensions...Present War* (Cmd. 6105), p. 9.
92 *Royal Warrant Concerning Retired Pay, Pensions and Other Grants for Members of the Military Forces Disabled, and of the Widows, Children and Dependants of Such Members Deceased in Consequence of the Present War, 1943* (Cmd. 6489), p. 37.
93 War Office, *Release and Resettlement: An Explanation of your Position and Rights* (London: HMSO, 1945), p. 44.
94 *Royal Warrant for the Retired Pay and Pensions...Present War* (Cmd. 6105), pp. 10, 15.
95 *Royal Warrant Concerning Retired Pay, Pension...Present War, 1943* (Cmd. 6489), p .38; *Ministry of Pensions: Improvements in War Pensions, 1945* (Cmd. 6714), p. 5.
96 *Royal Warrant for the Retired Pay and Pensions...Present War* (Cmd. 6105), p. 43.
97 *Royal Warrant for the Retired Pay and Pensions...Present War* (Cmd. 6105), p. 34.
98 *Royal Warrant Concerning the Retired Pay, Pensions and Other Grants for Members of the Military Forces and of the Nursing and Auxiliary Services*

Thereof Disabled, and for the Widows, Children, Parents and Other Dependents of Such Members Deceased in Consequence of Service after the 2nd September, 1939 (Cmd. 7699), p. 43.
99 Sandison, 'The medical services of the Ministry of Pensions', p. 226.
100 Ministry of Pensions, *War Pensions for Civilians and Members of the Civil Defence Services: Explanatory Notes on the Personal Injuries (Civilians) Scheme, 1944* (London: HMSO, 1944), p. 8.
101 Ministry of Pensions, *War Pensions for Civilians*, p. 13.
102 Hansard HC Deb., 24 October 1939, vol. 352, col. 1257; *Royal Warrant for the Retired Pay and Pensions...Present War* (Cmd. 6105), p. 10.
103 Ministry of Pensions, *War Pensions for Civilians*, p. 13; *Royal Warrant Concerning Retired Pay, Pension...Present War, 1943* (Cmd. 6489), p. 38.
104 Ministry of Pensions, *War Pensions for Civilians*, p. 17.
105 *Royal Warrant Concerning the Pay, Pensions and Other Grants in Cases where the Disablement or Death of Members of the Military Forces or the Home Guard is Due to Service During the Present War* (Cmd. 6558), p. 43.
106 Workmen (Compensation for Accidents). A Bill to Amend the Law with Respect to Workmen for Accidental Injuries Suffered in the Course of Their Employment, 1897 (Bill 213, 60 Vict.), p. 1.
107 Workmen (Compensation for Accidents) (Bill 213, 60 Vict.), p. 2; Bill to Consolidate and Amend the Law With Respect to Compensation to Workmen for Injuries Suffered in the Course of Their Employment (Employers' Liability: Workmen's Compensation Act (1897) Amendment (Seamen)), 1906 (Bill 123, 6 Edw. 7), pp. 9, 13.
108 Workmen (Compensation for Accidents) (Bill 213, 60 Vict.), pp. 4–5.
109 Bill (Passed Cap. 49) to Increase the Supplementary Allowances Payable to Workmen Entitled to Weekly Payments by Way of Compensation Under the Workmen's Compensation Act, 1945, and the Compensation Payable Under that Act on the Death of Workmen; and for Purposes Connected with the Matters Aforesaid (Workmen's Compensation (Temporary Increases)), 1932-43 (Bill 56, 6 and 7 Geo. 6), pp. 5–6. See also J. Bartrip, *Workmen's Compensation in Twentieth-Century Britain* (Aldershot: Avebury, 1987), p. 178.
110 *Royal Warrant Concerning Retired Pay, Pension...Present War, 1943* (Cmd. 6489), p. 38.
111 Bartrip, *Workmen's Compensation*, pp. 185–98.
112 *Social Insurance, Part II: Workmen's Compensation; Proposals for an Industrial Injury Allowance Scheme* (Cmd. 6551), p. 5.
113 *Social Insurance, Part II* (Cmd. 6551), p. 5.
114 Bartrip, *Workmen's Compensation*, p. 200.
115 *Social Insurance, Part II* (Cmd. 6551), p. 24; *Royal Warrant Concerning the Retired Pay, Pensions...after the 2nd September, 1939* (Cmd. 7699), p. 37.

116 Bill (Passed Cap. 62) to Substitute for the Workmen's Compensation Acts, a System of Insurance Against personal Injury Caused by accident Arising Out of and in the Cause of a Person's Employment and Against Prescribed Diseases and Injuries due to the Nature of a Person's Employment, and for Purposes Connected Therewith (as Amended by Standing Committee A on Re-Committal and on Report) (National Insurance (Industrial Injuries)), 1945–46 (Bill 79, 9 and 10 Geo. 6), pp. 4–7.
117 D.A. Gerber, 'Introduction: finding disabled veterans in history', in D.A. Gerber (ed.), *Disabled Veterans in History* (Ann Arbor: University of Michigan Press, 2000), pp. 11–14.
118 Bartrip, *Workmen's Compensation*, pp. 199–200.
119 Sandison, 'The medical services of the Ministry of Pensions', p. 194
120 'Ill health after recruitment', *British Medical Journal* (1 March 1941), 345.
121 'Tribunals for pensions claims', *Lancet* (9 May 1942), 576.
122 'The "ex"-man and his grievances', *Daily Mirror* (6 January 1941), p. 4.
123 *Changes in War Pensions* (Cmd. 6459), p. 2.
124 IWM SA, 20146, Roy Tomalin, reel 2.
125 IWM SA, 20146, Roy Tomalin, reel 2.
126 IWM SA, 17563, Norman Marshall, reel 3.
127 Scarry, *The Body in Pain*, pp. 14–15.

Conclusion

Between 1939 and 1945, millions of civilian men were recruited into the ranks of the British Army. They left their homes, families and jobs and had to adapt every part of their lives to the demands of military discipline and culture. This work has shown that the body was central to this process of becoming a soldier, and to the ways in which military life was organised, experienced and expressed.

What does this add to our understandings of the Second World War? For historians this work offers several valuable insights. As other accounts of the British Army in this period suggest, institutional arrangements were crucial to the ways in which the military authorities dealt with conscript soldiers, both at home and in the field. What we have seen is that these arrangements centred on the physical. It was not just a question of 'man management' but body management.[1] The army sorted men into bodily types which reflected the institution's cultural assumptions and sought to transform them through strict bodily regimes into figures that were imagined to be ideal. Examination procedures classified bodies according to the presence or absence of organic disease, the ability to see and hear and to undergo severe strain. Army training moulded and shaped bodies in order to instil the qualities needed for active service. Once in active operations, it was not just a question of achieving strategic advantages, superiority in weapons or deploying manpower most economically.[2] At the heart of the military operation was a determined effort to maintain and protect physical capacity through a range of interventions. These included inoculations, vaccinations, protective clothing and equipment and the segregation of bodies away from sources of disease.

Harrison is therefore right to suggest that a 'medical consciousness' was evident among the British authorities, whose careful monitoring of health contributed to military successes in the field.[3] In Europe, Africa

Conclusion

and the Middle East, officers and commanders advocated regimes of bodily discipline such as compulsory inspections and treatments, as well as education and propaganda, to keep their men free from disease. This work has shown that the control and surveillance of the soldier's body went beyond healthcare. From the moment that men entered the army, and even before that, their bodies were constantly watched, measured, evaluated and regulated, told when to eat, when to rest and how to move. Even in death, the body was controlled and organised by the state as men were confined to standardised graves. This was not simply to prevent the spread of disease, but to remove the corpse's symbolic presence as the reminder of men's own mortality.

As had been the case in the First World War, there was also a clear relationship between the military male body and masculinity. Between 1939 and 1945 dominant hegemonic conceptions of manliness affected both the values ascribed to the body and its actual shape and size.[4] Medical examiners looked for size, shape and musculature in the men that they were confronted with. Training regimes then invigorated the body, building it up and enhancing it in stamina, strength and endurance. As one instructor noted, 'it sorted the men out from the boys.'[5] Indeed, the Second World War was a period of continuity in many respects. It was a time when many of the issues that had concerned military, medical and government officials in former campaigns re-emerged. As it had in the First World War and the Boer conflict before it, in the Second World War the soldier's body became privileged as a barometer for measuring national strength. In an era dominated by concerns over the health of the British population, military medical examination results were viewed by some as evidence that the national stock had improved but by others that much still needed to be done. Anxieties about venereal disease also resurfaced during the Second World War when army leaders and medical staff were once again concerned about its effects on manpower. The authorities responded to the problem with a mix of traditional disciplinary methods and a new pragmatic approach. These reflected older notions about morality, masculinity and soldiering, as well as more modern conceptions of nationhood, citizenship and the simple recognition that not all men would abstain.

There were many other discourses informing official knowledge about the military male body, all of which represent important links between the army and civil society during these years.[6] For example, there were several intersections at which the military and industrial body met. This was particularly so during the pre-combat stages of army life.

Examination classified men's bodies in much the same way as industrial workers', as selection methods became more sophisticated in order to fit the right man to the right job. The wartime experiments conducted by the Military Personnel Research Committee also drew on ideas that had long been established in the sphere of industrial health research. Indeed, it was believed that when war ended, these processes and techniques could be reapplied into the industrial workplace. Thus, while in some respects the soldier's body was distinct or separate from the rest of society, in others it was just another form of 'working body', rendered productive through the same processes of rationalisation that were affecting other civilian men and women.

In the field of active service official perceptions of the body also drew upon older notions of race that were bound up with the politics of empire. Tropical climates, in particular, were viewed as incompatible with the soldier's constitution. This meant that troops either had to be physiologically adapted to, or protected from, the harmful effects of the hostile environment. The army's efforts to maintain health and efficiency among men serving overseas thus reflected two distinct conceptions of the human body. On the one hand, the body was perceived as open and in flux with its environment. On the other, it was a more fixed, biological entity that was limited in its capacity to adjust. These were ideas that had been prominent throughout the eighteenth and nineteenth centuries as part of the colonial mission. During the Second World War the British soldier's body was also constructed in marked opposition to dangerous 'other' bodies, particularly those of indigenous populations whose sanitary habits were targeted as the cause of disease. In this respect, the authorities again reflected older conceptions of biological difference and the Western civilising mission of the nineteenth century.

For those interested in ideas about the body, this work also offers several important insights. The army was clearly a point at which the state encountered the bodies of its citizens and was a world in which men's bodies were persistently under the control of many 'experts'. As such, it can be seen as part of the wider project of modernity. The regulation and surveillance of bodies was perhaps most extreme in the pre-combat stages of army life. At the examination the body was meticulously observed, measured in its functioning and broken down into its respective components so that doctors could isolate the faulty 'parts'. During training the body was internally and externally regulated: dressed, washed, fed, rested, confined to barracks and ordered for twenty-four hours a day through a timetable. Through an arduous routine of physical training, battle training and drill, instructors gradually transformed each body into a fit, ordered and replaceable cog in the wider military machine. Every recruit quickly learned that his role was to obey orders and to put the needs of the unit before his own. For those soldiers who came to participate in the army's experimentation programmes, their bodies were harmed, repaired and pushed to the limits of endurance so that the authorities could develop more effective fighting methods. These experiments often focused little on the health and wellbeing of the men but rather assessed their bodies according to their military capacities.

During the Second World War medical power also came to serve the interests of the state.[7] Civilian doctors and scientists sat on medical examination boards. They helped to develop more effective screening procedures and became involved in military human research. However, the designs of the state, military and medical authorities were not always the same and could frustrate each other. Some medical professionals and MPs raised opposition to physical selection standards and methods, which they believed to be inadequate. Rather than accepting new physical categories based on a man's specific abilities to see, hear, shoot or drive, they continued to conceptualise fitness in relation to the whole healthy physique. Some army officers were also dissatisfied with the quality of the men who arrived at their units, who had been passed as fit by medical examination boards but were found to be physically incapable of performing the duties required of their military roles. Medical experts too had their own specific agendas for soldiers' bodies. When it came to the army's trials of penicillin, civilian researchers were not necessarily driven by either the needs of the army or even a desire to cure the individual. Rather, they viewed the men whom they experimented on as useful scientific specimens in the pursuit of wider medical knowledge.

Yet there were limits to what the state, and medicine, could achieve. By looking not just at what was *done* to men's bodies but at what these bodies *did*, this work has demonstrated that the soldier was not simply a docile body, passively adjusting to the demands of the army, medicine or the state. Rather, the body was a real, material phenomenon that connected the soldier to his social world.[8] Oral histories have been particularly illuminating in this respect by providing access to the private lives of soldiers. They show that from enlistment to battle, soldiers were actively conscious of their bodies and the bodies of others. They experienced and used their bodies in particular ways, as they faced new obstacles and challenges and developed new attributes and skills. Clearly, bodies were not just shaped and moulded by the military authorities, but were subject to individual agency as men pursued their own desires and agendas.

No doubt, bodies were often far from ideal and served to frustrate the army's designs. This can be seen in the efforts made by many men to try and escape military duties. The most extreme or obvious acts of resistance included malingering, desertion, self-inflicted wounds and suicide. These behaviours occurred at every stage of the military journey. Men who did not want to enlist tried to fool their examiners with symptoms of illness or obtained false medical documentation. In training camps, recruits who wished to evade disliked activities or who simply 'wanted out' did the same, while those who longed to see their families at Christmas just went absent without leave. In the field of active operations, even the threat of imprisonment did not deter men from deserting or inflicting harm upon themselves. In some cases it offered a preferred alternative to the dangers of battle and the hardships of life on the front line.

Resistance could, however, be subtle and operate within the extant relations of power. As Silbey argues was the case in the First World War, medical examination between 1939 and 1945 was a highly negotiated gateway between the individual and the representatives of the state.[9] Men eager to serve tried to conceal their illnesses and conditions, while doctors sometimes let unfit men pass either to fill the ranks or simply because they knew their examinees. In one instance, a man who was examined by the family doctor was even allowed to decide his own physical classification. In this instance, both doctor and patient were able to manoeuvre around the exam's requirements.

Soldiers also found ways of satisfying their bodily needs without striving to escape from the military regime. Within their camps and depots men found opportunities to indulge in bodily pleasures and excesses. In their barrack rooms or during off-duty time, both officers and ordinary soldiers got drunk and had sex. Some even dressed as women and developed romantic relationships with each other. These carnivalesque performances may not simply have been about satisfying sexual desires but may have had wider symbolic meaning. In the field of active service, where formal discipline often became more relaxed, troops were also largely free to govern their own bodies. Those who chose to get drunk and have sex were not necessarily resisting the authority of their superiors, but were making the most of the new freedom that they had. Furthermore, men who indulged their bodily impulses were not always responding to the changes in lifestyle brought about by military service. There is no way of knowing, for example, if soldiers who were sexually promiscuous or who drank heavily were following patterns established in civilian life. It is equally possible that men who slept with each other had always been homosexual. In the field of active service, men who chose to be promiscuous or not to practise safe sex did so for sexual adventure, because they feared lessened pleasure or because they were drunk, lazy or simply did not know how to use the prophylactics supplied. These soldiers were not necessarily acting with wilful disobedience but were driven by various other factors.

Just as importantly, men who failed to live up to the army's expectations did not always intend to do so. There was the examinee who struggled to provide a urine sample, the recruit who could not run fast enough or march in step, or the soldier on active service who when faced with the enemy was literally paralysed with fear. Likewise, men who were eager to serve and who wished to be medically classified as A1 were worried that they might not reach the required standards. Roy Bolton at the start of this book, who was 'clumsy in a bodily sort of way', did not think that drill was 'very nice at all'.[10] Soldiers in combat tried desperately, but often without success, not to vomit and to control their shaking and shivering. Perhaps these stories can be read as evidence of the success of military authority by the fact that the men experienced their bodies as a source of frustration, embarrassment or constraint. The sense of disappointment expressed in many of their testimonies suggests that they were guided by a set of dispositions that were imposed from outside.[11] This does not, however, detract from the corporeal nature of the bodies, which were not – and indeed could not be – shaped as the army desired.

Soldiers also complied with the army's demands for their bodies in order to achieve productive ends of their own.[12] Recruits in training often came to enjoy the effects of military life as they grew bigger, fitter and stronger and gained greater access to food and medical attention. They engaged in their own physical transformations in order to make themselves look and feel good. Unlike many other histories of human experimentation, which argue that military personnel have no control over the matter of participation, this work has also shown that British soldiers were often willing volunteers. They took part in human trials in exchange for money, better meals or a break from normal duties. Other reasons to volunteer that the authorities may not have been aware of included the pleasures of competition, testing one's masculinity or the sense of simply doing one's duty. In such moments the body was empowering for the soldier as well as benefitting the state.

All of these instances show that the body was consistently at the foreground of the experience of army life and not just in times of pain or death.[13] For many men, the fully functioning body was a crucial concern, be it as symbol of masculinity or nationhood or as a material, sensuous entity that provided pleasure, desire and expectation. In active service, soldiers also came to experience pain, disease and fear – bodily conditions that might be termed as 'dysfunctional' – as a normal part of daily life.[14] In this respect, it might be, as Shilling suggests, more applicable to talk of the emergence of the healthy body as a process of 'reappearance' rather than 'disappearance'.[15] Indeed, this work has shown that men became acutely aware of their bodies when they became stronger and healthier. As such, the body, in pain, in death and in health was of fundamental importance to the soldier as well as for the military and medical officials who tried to control and adapt it.

To appreciate the significance of the body to the conduct of the Second World War therefore requires many layers of understanding. While military leaders, medical professionals and government officials may have tried to transform civilians into soldiers through their bodies and in some instances were able to achieve their objectives, in practice they could not always be successful. The body was difficult to control and to be controlled. Ultimately, it was an unstable object for the individual, the army and the wartime state.

Notes

1. Crang, *The British Army and the People's War*, p. 65.
2. French, *Raising Churchill's Army*, p. 11.
3. Harrison, *Medicine and Victory*, p. 278.
4. Bourke, *Dismembering the Male*; Carden-Coyne, *Reconstructing the Body*.
5. IWM SA, 11468, Ian Sinclair, reel 3.
6. Bourke, *Dismembering the Male*, p. 13.
7. Turner, *Regulating Bodies*, p. 47.
8. N. Crossley, 'Merleau-Ponty, the elusive body and carnal sociology', *Body and Society* 1:1 (1995), 19.
9. Silbey, 'Bodies and cultures collide', 65.
10. IWM SA, 23195, Roy Bolton, reel 2.
11. Goffman, *Behaviour in Public Places*, p. 35.
12. Frank, 'For a sociology of the body', p. 58.
13. Leder, *The Absent Body*, p. 84.
14. Leder, *The Absent Body*, p. 84.
15. Shilling, *The Body and Social Theory*, p. 187.

Bibliography

Archival sources

Imperial War Museums (IWM)

Sound Archive (IWM SA), oral history interviews

9732, Ronald Petts
9796, William Spearman
10601, Frederick Cottier
11207, James Bell
11468, Ian Sinclair
12239, Gerald Barnett
12240, Harold Atkins
12412, Joseph Day
12436, Ronald Sherlaw
12938, William Dunn
13128, James Ford
13230, Charles Bennett
13281, William Scroggie
13370, Arthur Thompson
14788, Albert Parker
14977, Ernest Harvey
14981, William Cornell
16084, Edward Kirby
16352, Frank Offiler
16714, Richard Forbes
17286, Douglas Arnold
17354, Dick Fiddament
17563, Norman Marshall
17628, Joseph Clark
17630, Eric Murray
17925, Stanley Shore
18255, Gordon Gent
18257, Charles Lord
18364, John Watts
18435, William Dilworth
18512, Russell King
18741, Rubin Wharmby
18743, Robert Ellison
19056, Ernest Lanning
19066, Arthur McCrystal
19638, Charles Jordan
19674, Basil Reeve
19802, Bill Scott
19804, Alexander Frederick
19805, Walter Chalmers
19867, John Buchanan
20009, Leslie Perry
20146, Roy Tomalin
20149, Alex Gilchrist
20201, George Cozens
20202, John Gray
20318, Charlie Workman
20373, Herbert Beddows
20620, William Tichard
20891, Joseph Inskip
20897, Bert Wheeler
21015, Bill Knights

Bibliography

21109, Leonard Harkins
21565, Bill Partridge
22072, Wilfred Hall
22075, Kenneth Bond
22195, Albert Rogers
22346, John Riggs
22387, Joe Stevens
22666, Henry Wilmott
23195, Roy Bolton

23216, William Corbould
23367, Henry Butterworth
24187, George Percy
24921, John Coldwell-Horstall
28618, Samuel Beard
30078, John Emerson
30407, Ralph Kirkton
30493, Ron Gray

Department of Documents
IWM Document Archive 125, private papers of R.H. Lloyd-Jones
IWM Document Archive, 2996, private papers of R.A. Graydon

Art and Popular Design Collection
Art.IWM PST 9063, 'Mosquitoes mean malaria: don't give them the chance'

Mass Observation Archive (MOA)

Topic Collections
TC29, Forces: Men in the Forces 1939–1956, 1/B, Observations: the public behaviour of men in the forces, 1939–40
TC29, Forces: Men in the Forces 1939–1956, 1/D, Recruitment, 1939
TC29, Forces: Men in the Forces 1939–1956, 1/F, The public and soldiers, 1940–41
TC29, Forces: Men in the Forces 1939–1956, 2/A, A. Calder Marshall
TC29, Forces: Men in the Forces 1939–1956, 2/B, Leonard England
TC29, Forces: Men in the Forces 1939–1956, 2/C, W.R. Lee
TC29, Forces: Men in the Forces 1939–1956, 2/D, Henry Novy
TC29, Forces: Men in the Forces 1939–1956, 2/E, Assorted short reports from other Observers in the Army
TC29, Forces: Men in the Forces 1939–1956, 3/A, National Service

File Reports
FR 836, An army depot in 1941, August 1941
FR 686, Education in the Armed Services, May 1941

Diaries
D 5006, Diary for June 1940
D 5039.9, Diary for February 1941
D 5061.1, Diary for January 1941
D 5061.1, Diary for February 1941
D 5061.1, Diary for April 1941
D 5134, Diary for December 1942

Bibliography

D 5165, Diary for November 1940
D 5165, Diary for December 1940
D 5175, Diary for May 1940
D 5188, Diary for May 1941
D. 5188, Diary for September 1941

The National Archives, Kew (TNA)

CAB21/914, Committee on the Work of Psychologists and Psychiatrists in the Services, 1940-42
FD1/7042, Military Personnel Research Committee: Rations Sub-Committee, 1941-42
FD1/6383, Military Personnel Research Committee: papers, 1940-47
FD1/7064, Sub-Committee on Analeptic Substances (Benzedrine), 1941-42
WO32/4726, MEDICAL: Committees (Code 18(c)): Report of the Horden Committee on the examination of recruits by civil medical boards under the Army Forces Act, 1940-47
WO32/4643, RECRUITING: General (Code 25 (A)): Physical standard required, 1 January 1936-31 December 1939
WO32/11550, MEDICAL: General (Code 18(A)): Psychiatric service in operational theatres, 1945
WO32/15773, Death penalty for desertion in the field: reintroduction, 1942
WO71/1074, General Courts Martial, Offence: Desertion, Lieutenant J. Speight, 1942
WO71/1114, General Courts Martial, Offence: Desertion, Lieutenant-Colonel S. Cook, 1945
WO71/10489, General Courts Martial, Offence: Indecency, C.B.A. Bernard, October 1940
WO177/1, Directorate of Medical Services, September 1939-June 1940
WO188/1449, Physiological observers (volunteers for experiments): employment; injury, 1940-41
WO189/1126, Effects on the skin when drops of HT (mustard gas variant) 60/40 and drops of HL (mustard-lewisite variant) 50/50 are dropped on subjects in winter service dress, 1 January 1939-31 December 1939
WO189/2270, Medical report on casualties produced by airburst mustard gas shell, 1 January 1942-31 December 1942
WO189/2293, A physiological test for the harassing effect of lachrymators on vision, and a suggested method of using the rangefinder in gas chamber trials, 1 January 1942-31 December 1942
WO189/2321, The use of an assault course in the assessment of the arsenical smokes, 1 January 1942-31 December 1942
WO189/2848, History of the Service Volunteer Observer Scheme at Chemical Defence Experimental Establishment (CDEE), November 1959

Bibliography

WO203/691, Use of Benzedrine by armoured troops: report, December 1942
WO222/27, Health of the Army, statement by Director of Hygiene to Army Hygiene Advisory Committee, 1941
WO222/97, Benzedrine in war operations: notes from the Middle East, 1942
WO222/1302, Consulting venereolgist, Middle East Force, July 1940–December 1943
WO222/1543, 1 Traumatic Shock Team, October 1944–March 1945
WO252/1316, Report on Southern Italy: climate and medical, June 1943
WO277/7, Brigadier A.B. MacPherson, Discipline 1939–1945
WO277/12, Manpower problems, 1939–45, 1 January 1949–31 December 1949
WO286/11, Volunteers for physiological tests at Chemical Defence Experimental Station, Porton, April 1952–June 1960
WO293/25, Army Council: Instructions, 1 January 1940–31 December 1940
TS27/398, Medical Research Council: Question of the legal position of the council if volunteer subjects for proposed experiments should die, 1933; 1945

Wellcome Archives, London

RAMC/349, A 'malingerer's guide' to how to appear ill disguised as a book of matches, one of the many distributed among British troops in Italy towards the end of the Second World War, 1944
RAMC 466/48, Reports, statistics and posters re venereal disease amongst allied troops in Italy, 1943–44
RAMC/1129, Report on Army Physical Development Centres and handbooks for Army Physical Training Centre instructors, 1943–45
GC135/B.1/3, Service psychiatry monographs: Report no. 47, Lt.-Col. MacKeith, Some comments on the VD problem in an expeditionary force, 1944
Wellcome Images L0023970, 'Just another "Claptrap". You can ruin your future with – V.D!' Poster designed by Stacey Hopper to warn Allied troops in Italy about the dangers of venereal disease, 1943–44 http://images.wellcome.ac.uk (accessed January 2012)

Published sources

Parliamentary sources

Command papers
Ministry of Labour and National Service, *Report No 1 Upon the Physical Examination of Men of Military Age by National Service Boards from Nov 1st 1917–October 31st* [Cmd. 504]
Royal Warrant for the Retired Pay and Pensions etc. of Members of the Military Forces Disabled, and of the Widows, Children and Dependants of Such Members Deceased in Consequence of the Present War [Cmd. 6105]

Bibliography

Committee of the Privy Council for Medical Research, *Report of the Medical Research Council for the Year 1938-1939* (Cmd. 6163)

Report of the Medical Advisory Committee on the Use of Mass Miniature Radiography in the Detection of Pulmonary Tuberculosis among Recruits for H. M. Forces, 1942 (Cmd. 6353)

Summary Report by the Ministry of Health for the Period from 1st April, 1941 to 31st March, 1942 (Cmd. 6394)

Changes in War Pensions, 1942-43 (Cmd. 6459)

Royal Warrant Concerning Retired Pay, Pensions and Other Grants for Members of the Military Forces Disabled, and of the Widows, Children and Dependants of Such Members Deceased in Consequence of the Present War, 1943 (Cmd. 6489)

Social Insurance, Part II: Workmen's Compensation; Proposals for an Industrial Injury Allowance Scheme, 1944 (Cmd. 6551)

Royal Warrant Concerning the Pay, Pensions and Other Grants in Cases Where the Disablement or Death of Members of the Military Forces or the Home Guard is due to Service During the Present War, 1944 (Cmd. 6558)

Ministry of Pensions: Improvements in War Pensions, 1945 (Cmd. 6714)

Strength and Casualties of the Armed Forces and Auxiliary Services of the United Kingdom 1939-1945 (Cmd. 6832)

Pay and Allowances of the Armed Forces, 1941-42 (Cmd. 6385)

Report of the Inter-Departmental Committee on the Assessment of Disablement due to Specified Injuries, 1947 (Cmd. 7076)

Ministry of Labour and National Service, *Report for the Years 1939-1946* (Cmd. 7225)

Medical Research in War: Report of the Medical Research Council for the Years 1939-1945 (Cmd. 7335)

Royal Warrant Concerning the Retired Pay, Pensions and Other Grants for Members of the Military Forces and of the Nursing and Auxiliary Services Thereof Disabled, and for the Widows, Children, Parents and Other Dependants of Such Members Deceased in Consequence of Service after the 2nd September, 1939 (Cmd. 7699)

Workmen (Compensation for Accidents). A Bill to Amend the Law with Respect to Workmen for Accidental Injuries Suffered in the Course of Their Employment, 1897 [Bill 213, 60 Vict.]

Statutes

Bill to Consolidate and Amend the Law With Respect to Compensation to Workmen for Injuries Suffered in the Course of Their Employment (Employers' Liability: Workmen's Compensation Act (1897) Amendment (Seamen)) 1906 (Bill 123 6 Edw. 7)

Bibliography

Workmen (Compensation for Accidents). A Bill to Amend the Law with Respect to Workmen for Accidental Injuries Suffered in the Course of Their Employment, 1897 (Bill 213, 60 Vict.)

Bill (Passed Cap. 49) to Increase the Supplementary Allowances Payable to Workmen Entitled to Weekly Payments by Way of Compensation Under the Workmen's Compensation Act, 1945, and the Compensation Payable Under that Act on the Death of Workmen; and for Purposes Connected with the Matters Aforesaid (Workmen's Compensation (Temporary Increases)) (Bill 56, 6 and 7 Geo. 6)

Bill (Passed Cap. 62) to Substitute for the Workmen's Compensation Acts, a System of Insurance Against Personal Injury Caused by Accident Arising out of and in the Cause of a Person's Employment and Against Prescribed Diseases and Injuries dues to the Nature of a Person's Employment, and for Purposes Connected Therewith (as Amended by Standing Committee A on Re-Committal and on Report) (National Insurance (Industrial Injuries)) (Bill 79, 9 and 10 Geo. 6)

Hansard Parliamentary Debates

Hansard HC Deb., 22 June 1939, vol. 348, cols 2427–34
Hansard HC Deb., 17 October 1939, vol. 352, col. 712
Hansard HC Deb., 19 October 1939, vol. 352, cols 1056–7
Hansard HC Deb., 24 October 1939, vol. 352, cols 1253–323
Hansard HC Deb., 7 December 1939, vol. 355, cols 787–8
Hansard HC Deb., 27 February 1940, vol. 357, col. 1910
Hansard HC Deb., 14 March 1940, vol. 358, cols 1447–8
Hansard HC Deb., 17 April 1940, vol. 359, col. 180
Hansard HC Deb., 5 November 1940, vol. 365, col. 1928
Hansard HC Deb., 2 December 1941, vol. 376, col. 1030
Hansard HC Deb., 29 April 1942, vol. 379, col. 1044.
Hansard HC Deb., 9 September 1942, vol. 385, cols 1558–60
Hansard HC Deb., 16 March 1944, vol. 398, col. 375

Published memoirs

Coutts, B., *A Scotsman's War* (Edinburgh: Mercat Press, 1995)
Elliot, W.A., *Esprit de Corps: A Scots Guards Officer on Active Service, 1943–1945* (Wimborne: Michael Russell, 1996)
Grant, P., *A Highlander Goes to War: A Memoir, 1939–1946* (Edinburgh: Pentland Press, 1995)
MacDonald Fraser, G., *Quartered Safe out Here: A Recollection of the War in Burma* (London: Harvill, 1992)
McCallum, J., *Journey with a Pistol* (London: Victor Gollanz, 1959)
Milligan, S., *Adolph Hitler, My Part in his Downfall* (London: Penguin, 1971)
Watts, J.C., *Surgeon at War* (London: Allen and Unwin, 1955)

Bibliography

Wingfield, R.M., *The Only Way Out: An Infantryman's Autobiography of the North-West Europe Campaign, August 1944–February 1945* (London: Hutchinson, 1955)

Newspapers and periodicals

American Journal of Psychology
British Journal of Industrial Medicine
British Journal of Medical Psychology
British Medical Bulletin
British Medical Journal
Canadian Medical Association Journal
Journal of Mental Science
Journal of Physiology
Journal of the Royal Army Medical Corps
Lancet
Manchester Guardian
Mirror
Observer
Scotsman
The Times

Contemporary publications

'Anomaly', *The Invert and his Social Adjustment* (London: Bailliere, Tindahl and Cox, 1927)
Bassett Jones, A., L.J. Llewellyn and W.M. Beaumont, *Malingering: Or the Simulation of Disease* (London: William Heinemann, 1917)
Burje, E.T., *Tropical Tips for Troops in the Tropics or How to Keep Fit in the Tropics* (London: Heinemann, 1942)
Central Statistical Office, *Statistical Digest of the War* (London: HMSO, 1951)
Collie, J., *Malingering and Feigned Sickness* (London: Edward Arnold, 1913)
Crew, F.A.E., *The Army Medical Services: Administration, Volume I* (London: HMSO, 1953)
Crew, F.A.E., *The Army Medical Services: Administration, Volume II* (London: HMSO, 1953)
Franklin Mellor, W. (ed.), *Medical History of the Second World War: Casualties and Medical Statistics* (London: HMSO, 1972)
General Headquarters, Middle East Forces, *Psychiatric Casualties: Hints to Medical Officers in the Middle East Forces* (Cairo: Middle East Forces, 1942)
Green, F.H.K. and Sir Gordon Kovell, *Medical Research* (London: HMSO, 1953)
Henderson, D.K., 'The significance of fear', *Edinburgh Postgraduate Lectures in Medicine* 2 (1941), 27–39

Bibliography

Hirschfeld, M., *The Sexual History of the World War* (New York: Panurge Press, 1934)

Legros, L.A. and H.C. Weston, 'On the design of machinery in relation to the operator', *Industrial Fatigue Research Board, Report No. 36* (London: HMSO, 1926)

MacNalty, Sir Arthur Salusbury (ed.), *The Civilian Health and Medical Services: Volume I, The Ministry of Health Services; Other Civilian Health and Medical Services* (London: HMSO, 1953)

MacNalty, Sir Arthur Salusbury and W. Franklin Mellor, *The Medical Services in War: The Principal Medical Lessons of the Second World War: Based on the Official Medical Histories of the United Kingdom, Canada, Australia, New Zealand and India* (London, HMSO, 1968)

Martin, W.J., 'The physique of young adult males', *Medical Research Council Memorandum 20* (London, HMSO, 1949)

Ministry of Labour and National Service, *Manpower: The Story of Britain's Mobilisation for War* (London: HMSO, 1944)

Ministry of Pensions, *War Pensions for Civilians and Members of the Civil Defence Services: Explanatory Notes on the Personal Injuries (Civilians) Scheme, 1944* (London: HMSO, 1944)

Myers, C.S., 'Introduction', in C.S. Myers (ed.), *Industrial Psychology* (London: HMSO, 1943), pp. 7-15

Myers, C.S. (ed.), *Industrial Psychology* (London: HMSO, 1943)

Reports of the Industrial Fatigue Research Board, No. 16: Three Studies in Vocational Selection (General Series No. 6)(London: HMSO, 1922)

Sandison, A., 'The medical services of the Ministry of Pensions', in Sir A. Salusbury MacNalty and W. Franklin Mellor (eds.), *The Civilian Health and Medical Services: Volume II* (London: HMSO, 1955), pp. 139-26

Taylor, F.W., *The Principles of Scientific Management* (New York and London: Harper, 1911)

War Office, *Basic and Battle Physical Training, Part I: General Principles of Basic and Battle Physical Training and Methods of Instruction* (London: HMSO, 1944)

War Office, *Basic and Battle Physical Training, Part II: Basic Physical Training Tables and Basic Physical Efficiency Tests* (London: HMSO, 1944)

War Office, *Field Service Regulations, Volume I: Organization and Administration* (London: HMSO, 1930)

War Office, *Handbook of Military Hygiene* (London: HMSO, 1943)

War Office, *Handbook of Military Law* (London: HMSO, 1943; first published 1929, reprinted 1939)

War Office, *Manual of Military Cooking and Dietary: Part 1, General* (London: HMSO, 1940)

War Office, *Physical and Recreational Training* (London: HMSO, 1941)

Bibliography

War Office, *Release and Resettlement: An Explanation of your Position and Rights* (London: HMSO, 1945)
War Office, *Statistics of the Military Effort of British Empire during the Great War* (London: HMSO, 1922).
War Office, *The Soldier's Welfare: Notes for Officers* (London: HMSO, 1941)
War Office, General Headquarters, Middle East Forces, *Psychiatric Casualties: Hints to Medical Officers in the Middle East Forces* (London: HMSO, 1942)
Whitting, R.E., 'Civilian medical recruiting boards', in Sir Arthur Salusbury MacNaulty (ed.), *The Civilian Health and Medical Services: Volume 1, The Ministry of Health Services; Other Civilian Health and Medical Services* (London: HMSO, 1953) pp. 346–66

Secondary sources

Addison, P. and A. Calder, *Time to Kill: The Soldier's Experience of War in the West* (London: Pimlico, 1997)
Anderson, J.A., *War, Disability and Rehabilitation in Britain: 'Soul of a Nation'* (Manchester: Manchester University Press, 2011)
Armitage, S.H. and S. Berger Gluck, 'Reflection on women's oral history: an exchange', in R. Perks and A. Thomson (eds.), *The Oral History Reader* (London: Routledge, 2nd edn, 2006), pp. 73–82
Armstrong, N. and E. Murphy, 'Conceptualising resistance', *Health* (2011), 1–13
Balchin, W.G.V., 'United Kingdom geographers in the Second World War: a report', *Geographical Journal* 53:2 (1987), 159–80
Bartrip, P.W.J., *Workmen's Compensation in Twentieth-Century Britain* (Aldershot: Avebury, 1987)
Beevor, A., *D-Day: The Battle for Normandy* (London: Penguin, 2010)
Bendelow, G. and S. Williams, 'Pain and the mind–body dualism: a sociological approach', *Body and Society* 1:2 (1995), 83–103
Berthlot, J.M., 'Sociological discourse and the body', in M. Featherstone, M. Hepworth and B. Turner (eds.), *The Body: Social Process and Cultural Theory* (London: Sage, 1996), pp. 390–404
Bérubé, A., *Coming Out under Fire: The History of Gay Men and Women in World War Two* (New York: Free Press, 2000)
Black, L., 'Health Law: Informed consent in the military: the anthrax vaccination', *American Medical Association Journal of Ethics*, 9:10 (2007), 698–702
Bland, L., '"Guardians of the race" or "vampires upon the nation's health"? Female sexuality and its regulation in early twentieth-century Britain', in D. Leonard and E. Whitelegg (eds.), *The Changing Experience of Women* (Oxford: Martin Robertson in association with the Open University, 1982), pp. 375–88
Bornat, J., R. Perks, P. Thompson and J. Walmsley (eds.), *Oral History, Health and Welfare* (London: Routledge, 2000)

Bibliography

Bourdieu, P., *Distinction: A Social Critique of the Judgement of Taste* (London: Routledge, 1984)

Bourke, J., *An Intimate History of Killing: Face-to-Face Killing in Twentieth-Century Warfare* (London: Granta, 1999)

Bourke, J., *Dismembering the Male: Men's Bodies, Britain and the Great War* (London: Reaktion, 1996)

Bourke, J., *Fear: A Cultural History* (London: Virago, 2005)

Braverman, H., *Labor and Monopoly Capital: The Degradation of Work in the Twentieth Century* (New York: Monthly Review Press, 1974)

Bullington, J., 'Body and self: a phenomenological study on the ageing body and identity', *Medical Humanities* 32 (2006), 25–31

Burnett, K.A. and M. Holmes, 'Bodies, battlefields and biographies: scars and the construction of the body as heritage', in S. Cunningham-Burley and K. Backett-Milburn (eds.), *Exploring the Body* (Basingstoke: Palgrave, 2001), pp. 21–36

Bynum, W., 'Reflections on the history of human experimentation', in F. Spicker, I. Alon, A. De Vries and H. Tristam Engelhardt Jr (eds.), *The Use of Human Beings in Research: With Special Reference to Clinical Trials* (Dordrecht: Kluwer Academic Publications, 1988), pp. 29–46

Caforio, G., *Handbook of the Sociology of the Military* (Kluwer Academic: New York, 2003)

Canning, K., 'Feminist history after the linguistic turn: historicizing discourse and experience', *Signs: Journal of Women in Culture and Society* 19:2 (1994), 368–404

Carden-Coyne, A., *Reconstructing the Body, Classicism, Modernism and the First World War* (Oxford: Oxford University Press, 2009)

Chandler, D. and I. Beckett (eds.), *The Oxford Illustrated History of the British Army* (Oxford: Oxford University Press, 1994)

Clark, K. and M. Holquist, *Mikhail Bakhtin* (Cambridge, MA: Harvard University Press, 1986)

Collingham, E.M., *Imperial Bodies: The Physical Experience of the Raj, c.1800–1947* (Cambridge: Polity, 2001)

Connell, R.W., *Masculinities* (Cambridge: Polity, 2nd edn, 1995)

Connell, R.W., *The Men and the Boys* (Cambridge: Polity, 2000)

Constantine, S., *Social Conditions in Britain, 1918–1939* (London: Methuen, 1983)

Cooter, R., 'Malingering in modernity: psychological scripts and adversarial encounters during the First World War', in R. Cooter, M. Harrison and S. Sturdy (eds.), *Medicine and Modern Warfare* (Atlanta, GA: Rodopi, 1999), pp. 125–48

Cooter, R., 'Medicine and the goodness of war', *Canadian Bulletin of Medical History* 12 (1990), 147–59

Bibliography

Cooter, R., M. Harrison and S. Sturdy (eds.), *Medicine and Modern Warfare* (Atlanta, GA: Rodopi, 1999)
Coveney, J., 'The government and ethics of health promotion: the importance of Michel Foucault', *Health Education Research, Theory and Practice* 13:3 (1998), 459–68.
Crang, J., 'Square pegs in round holes: other rank selection in the British Army, 1939–45', *Journal of the Society for Army Historical Research* 77 (1999), 293–8
Crang, J., *The British Army and the People's War, 1939–1945* (Manchester: Manchester University Press, 2000)
Crang, J., 'The British Army as a social institution', in H. Strachan (ed.), *The British Army: Manpower and Society into the Twenty-First Century* (London: Frank Cass, 2000)
Crossley, N., 'Body-subject/body-power, agency, inscription and control in Foucault and Merleau-Ponty', *Body and Society* 2:2 (1996), 99–116
Crossley, N., 'Merleau-Ponty, the elusive body and carnal sociology', *Body and Society* 1:1 (1995), 17–46
Cunningham-Burley, S. and K. Backett-Milburn (eds.), *Exploring the Body* (Basingstoke: Palgrave, 2001)
Davidson, R., '"Searching for Mary, Glasgow": contact tracing for sexually transmitted diseases in twentieth-century Scotland', *Social History of Medicine* 9:2 (1996), 195–214
Davis, J.B. (ed.), *Grand Campaigns of World War II* (Leicester: Silverdale Books, 2002)
Dawson, G., *Soldier Heroes: British Adventure, Empire and the Imagining of Masculinities* (London: Routledge, 1994)
Delaforce, P., *Churchill's Desert Rats in West Africa and Italy* (Barnsley: Pen and Sword Military, 2009)
Devine, J., *Forgotten Voices of Dunkirk* (London: Ebury Press, 2010)
Doherty, R., *Ubique: The Royal Artillery in the Second World War* (Stroud: History Press, 2008)
Douglas, M., *Purity and Danger* (London: Routledge, 1966)
Durkheim, E., *The Elementary Forms of the Religious Life* (London: Allen and Unwin, 1976)
Ellis, H., *Studies in the Psychology of Sex, Volume 2: Sexual Inversion* (New York: Random House, 1937)
Ellis, J., *The Sharp End: The Fighting Man in World War II* (London: Aurum Press, 2009)
Evans, R., *Gassed: British Chemical Warfare Experiments on Humans at Porton Down* (London: House of Stratus, 2000)
Featherstone, M., M. Hepworth and B. Turner (eds.), *The Body: Social Process and Cultural Theory* (London: Sage, 1991)
Ferris, P., *Sex and the British: A Twentieth-Century History* (London: Penguin, 1993)

Bibliography

Fischer-Tine, H. and M. Mann (eds.), *Colonialism as Civilizing Mission: Cultural Ideology in British India* (London: Anthem Press, 2004)

Forth, C., *Masculinity in the Modern West, Gender, Civilization and the Body* (Basingstoke: Palgrave, 2008)

Forty, G., *British Army Handbook, 1939-1945* (London: Chancellor Press, 2000)

Foucault, M., *Discipline and Punish: The Birth of the Prison* (London: Penguin, 1979)

Foucault, M., 'Governmentality', in G. Burchell, C. Gordon and P. Miller (eds.), *The Foucault Effect: Studies in Governmentality* (Chicago: University of Chicago Press, 1991), pp. 87-105

Foucault, M. 'Technologies of the self', in L.H. Martin and H. Gutman (eds.), *Technologies of the Self: A Seminar with Michel Foucault* (Amherst, MA: University of Massachusetts Press, 1988), pp. 16-49

Foucault, M., *The History of Sexuality: Volume 1, The Will to Knowledge* (London: Penguin, 1978)

Francis, M., *The Flyer: British Culture and the Royal Air Force, 1939-1945* (Oxford: Oxford University Press, 2008)

Frank, A., 'For a sociology of the body: an analytical review', in M. Featherstone, M. Hepworth and B. Turner (eds.), *The Body: Social Processes and Cultural Theory* (London: Sage, 1991), pp. 36-102

Fraser, D., *And we Shall Shock them: The British Army in the Second World War* (London: Hodder and Stoughton, 1983)

Freeden, M., 'Eugenics and progressive thought: a study in ideological affinity', *Historical Journal* 22 (1979), 645-71.

French, D., 'Discipline and the death penalty in the British Army in the war against Germany during the Second World War', *Journal of Comparative History* 33:4 (1998), 531-45

French, D., *Raising Churchill's Army: The British Army and the War against Germany, 1919-1945* (Oxford: Oxford University Press, 2000)

French, D., 'Tommy is no soldier: the morale of the Second British Army in Normandy, June-August 1944', *Journal of Strategic Studies* 19:4 (1996), 154-78

Frisna, M.E., 'Medical ethics in military biomedical research', in T.E. Beam and L.R. Sparacino (eds.), *Military Medical Ethics, Volume II* (Falls Church, VA: Office of the Surgeon General, Department of the Army, United States of America, 2003), pp. 533-63

Fussell, P., *Wartime: Understanding and Behaviour in the Second World War* (Oxford: Oxford University Press, 1999)

Garber, M., *Vested Interests: Cross-Dressing and Cultural Anxiety* (London: Routledge, 1992)

Gerber, D., 'Disabled veterans, the state, and the experience of disability in Western societies, 1914-1950', *Journal of Social History* 36:4 (2003), 899-916

Bibliography

Gerber, D.A., 'Introduction: finding disabled veterans in history', in D.A. Gerber (ed.) *Disabled Veterans in History* (Ann Arbor: University of Michigan Press, 2000), pp. 1–54

Gerber, D.A. (ed.), *Disabled Veterans in History* (Ann Arbor: University of Michigan Press, 2000)

Gerth, H.H. and C.W. Mills (eds.), *From Max Weber: Essays in Sociology* (London: Routledge, 1948)

Gilbert, A.N., 'Law and honour among eighteenth-century British Army officers', *Historical Journal* 19:1 (1976), 75–87

Gill, R., K. Henwood and C. McLean, 'Body projects and the regulation of normative masculinity', *Body and Society* 11:1 (2005), 37–62

Gilman, S.L., *Health and Illness: Images of Difference* (London: Reaktion, 1995)

Gledhill, C. and G. Swanson, *Nationalising Femininity: Culture, Sexuality, and British Cinema in the Second World War* (Manchester: Manchester University Press, 1996)

Goffman, E., *Behaviour in Public Places: Notes on the Social Organization of Gatherings* (New York: Free Press, 1963)

Goffman, E., *Stigma: Notes on the Management of Spoiled Identity* (Englewood Cliffs, NJ: Prentice-Hall, 1963)

Goodman, J., A. McElligott and L. Marks, 'Making human bodies useful: historicizing medical experiments in the twentieth century', in, J. Goodman, A. McElligott and L. Marks (eds.), *Useful Bodies: Humans in the Service of Medical Science in the Twentieth Century* (Baltimore: Johns Hopkins University Press, 2003), pp. 1–23

Goodman, J., A. McElligott and L. Marks (eds.), *Useful Bodies: Humans in the Service of Medical Science in the Twentieth Century* (Baltimore: Johns Hopkins University Press, 2003)

Gordon, C. and P. Miller (eds.), *The Foucault Effect: Studies in Governmentality* (Chicago: University of Chicago Press, 1991)

Graham, D., *Against Odds: Reflections on the Experiences of the British Army, 1914–45* (Basingstoke: Macmillan, 1999)

Hall, L., *Sex, Gender and Social Change in Britain since 1880* (Basingstoke: Macmillan, 2000)

Hallam, E., J. Hockey and G. Howarth, *Beyond the Body: Death and Social Identity* (London: Routledge, 1999)

Hallam, E., J. Hockey and G. Howarth, 'The body in death', in N. Watson and S. Cunningham-Burley (eds.), *Reframing the Body* (Basingstoke: Palgrave, 2001), pp. 63–77

Hancock, P., B. Hughes, E. Jagger, K. Paterson, R. Russell, E. Tuelle-Winton and M. Tyler (eds.), *The Body, Culture and Society: An Introduction* (Buckingham: Open University Press, 2000).

Harrison, M., *Climates and Constitutions: Health, Race, Environment and British Imperialism in India, 1600–1850* (Oxford: Oxford University Press, 1999)

Bibliography

Harrison, M., 'Medicine and the culture of command: the case of malaria control in the British Army during the two World wars', *Medical History* 40 (1996), 437-52

Harrison, M., *Medicine and Victory: British Military Medicine in the Second World War* (Oxford: Oxford University Press, 2004).

Harrison, M., 'Sex and the citizen soldier: health, morals and discipline in the British army during the Second World War', in R. Cooter, M. Harrison and S. Sturdy (eds.), *Medicine and Modern Warfare* (Atlanta, GA: Rodopi, 1999), pp. 225-49

Harrison, M., 'The British Army and the problem of venereal disease in France and Egypt during the First World War', *Medical History* 39 (1995), 133-58.

Harrison, M., 'The tender frame of man: disease, climate and racial difference in India and the West Indies', *Bulletin of the History of Medicine* 70:1 (1996), 68-93

Harrison-Place, T., *Military Training in the British Army, 1940-1944: From Dunkirk to D-Day* (London: Frank Cass, 2000)

Higate, P.R., 'The body resists: everyday clerking and unmilitary practice', in S. Nettleton and J. Watson (eds.), *The Body in Everyday Life* (London: Routledge, 1998), pp. 180-98

Higate, P.R. (ed.), *Military Masculinities, Identity and the State* (Westport, CT: Praeger, 2003)

Hockey, J., 'Head down, bergen on, mind in neutral: the infantry body', *Journal of Political and Military Sociology* 30:1 (2002), 148-71

Hollander, J.A. and R.L. Einwohner, 'Conceptualizing resistance', *Sociological Forum* 19 (2004), 539-42

Hopton, J., 'The state and military masculinity', in P.R. Higate (ed.), *Military Masculinities: Identity and the State* (Westport, CT: Praeger, 2003), pp. 111-23

Houlbrook, M., *Queer London: Pleasures and Perils in the Sexual Metropolis* (Chicago: University of Chicago Press, 2006)

Howson, A., *The Body in Society: An Introduction* (Cambridge: Polity, 2004)

Hughes, B., 'Medicalized bodies', in P. Hancock, B. Hughes, E. Jagger, K. Paterson, R. Russell, E. Tuelle-Winton and M. Tyler (eds.), *The Body, Culture and Society: An Introduction* (Buckingham: Open University Press, 2000), pp. 12-28

Jarvis, C.S., *The Male Body at War: American Masculinity during World War II* (Chicago: Northern Illinois University Press, 2004)

Jones, E. and N.T. Fear, 'Alcohol use and misuse within the military: a review', *International Review of Psychiatry* 23 (2011), 166-72

Jones, G., 'Eugenics and social policy between the wars', *Historical Journal* 25:3 (1982), 717-28

Jones, G., *Social Hygiene in Twentieth-Century Britain* (London: Croom Helm, 1986)

Bibliography

Keegan, J., *Churchill's Generals* (London: Warner, 1992)
Keegan, J., *The Second World War* (London: Pimlico, 1997)
Kreis, S., 'The diffusion of scientific management: the Bedaux Company in America and Britain, 1926-1945', in S. Kreis (ed.), *A Mental Revolution: Scientific Management since Taylor* (Columbus: Ohio State University Press, 1992), pp. 156-74
Lant, A., *Blackout: Reinventing Women for Wartime British Cinema* (Princeton: Princeton University Press, 1991)
Leder, D., *The Absent Body* (Chicago: Chicago University Press, 1990)
Leder, D. (ed.), *The Body in Medical Thought and Practice* (London: Kluwer Academic, 1992)
Lederer, S., 'Military personnel as research subjects', in S.G. Post (ed.), *Encyclopedia of Bioethics* (London: Macmillan Reference, 3rd edn, 2004), pp. 1843-6
Lederer, S., *Subjected to Science: Human Experimentation in America before the Second World War* (Baltimore: Johns Hopkins University Press, 1995)
Lee, N. de, 'Oral history and British soldiers' experience', in P. Addison and A. Calder (eds.), *Time to Kill: The Soldier's Experience of War in the West* (London: Pimlico, 1997), pp. 360-8
Levine, P., *Prostitution, Race and Politics: Policing Venereal Disease in the British Empire* (London: Routledge, 2003)
Littler, C.R., *The Development of the Labour Process in Capitalist Societies: A Comparative Study of the Transformation of Work Organisation in Britain, Japan and the USA* (London: Heinemann Educational, 1982)
Long, V., *The Rise and Fall of the Healthy Factory: The Politics of Industrial Health in Britain, 1914-1960* (London: Palgrave, 2010)
Longden, S., *To the Victor the Spoils: Soldiers Lives from D-Day to VE-Day* (London: Robinson, 2007)
McCallum, J.E., *Military Medicine from Ancient Times to the 21st Century* (Santa Barbara: ABC-Clio, 2008)
McIvor, A.J., *A History of Work in Britain, 1880-1950* (London: Palgrave, 2001)
McLynn, F., *The Burma Campaign: Disaster into Triumph, 1942-45* (London: Bodley Head, 2010)
McManus, J., S.G. Mehta, A.R. McClinton, R.A. De Lorenzo and T.W. Baskin, 'Informed consent and ethical issues in military medical research', *Academic Emergency Medicine*, 12:11 (2005), 1120-6
Mangan, J.A. and J. Walvin, 'Introduction', in J.A. Mangan, and J. Walvin (eds.), *Manliness and Morality, Middle-class Masculinity in Britain and America, 1800-1940* (New York: St Martin's Press, 1987), pp. 1-6
Mangan, J.A. and J. Walvin (eds.), *Manliness and Morality, Middle-Class Masculinity in Britain and America, 1800-1940* (New York: St Martin's Press, 1987)

Bibliography

Martin, L.H. and H. Gutman (eds.), *Technologies of the Self: A Seminar with Michel Foucault* (Amherst, MA: University of Massachusetts Press, 1988).

Mason, T. and E. Riedi, *Sport and the Military: The British Armed Forces, 1880–1960* (Cambridge: Cambridge University Press, 2010)

Mauss, M., 'Techniques of the body', *Economy and Society* 2:1 (1973), 70–88

May, C., 'The clinical encounter and the problem of context', *Sociology* 41:1 (2007), 29–45

Mayhew, M., 'The 1930s nutrition controversy', *Journal of Contemporary History* 23 (1988), 445–56

Mellanby, K., *Human Guinea Pigs* (London: Merlin, 1973)

Merleau-Ponty, M., *The Phenomenology of Perception*, trans. Colin Smith (London: Routledge, 2nd edn, 2002)

Moreno, J.D., *Undue Risk: Secret State Experiments on Humans* (London: Routledge, 2000)

Morgan, D., 'Theater of war: combat, the military and masculinities', in H. Brod and M. Kaufman (eds.), *Theorising Masculinities* (London: Sage, 1994), pp. 165–82

Mosse, G., *The Image of Man: The Creation of Modern Masculinity* (Oxford: Oxford University Press, 1996)

Narvaez, R.F., 'Embodiment, collective memory and time', *Body and Society* 12:3 (2006), 51–72

Nash, L., 'Finishing nature: harmonizing bodies and environments in late nineteenth-century California', *Environmental History* 8:1 (2003), 25–52

Nettleton, S. and J. Watson, 'The body in everyday life: an introduction', in S. Nettleton and J. Watson (eds.) *The Body in Everyday Life* (London: Routledge, 1998), pp. 1–24

Nettleton, S. and J. Watson (eds.), *The Body in Everyday Life* (London: Routledge, 1998)

Neushul, P., 'Fighting research: army participation in the clinical testing and mass production of penicillin during the Second World War', in R. Cooter, M. Harrison and S. Sturdy (eds.), *War, Medicine and Modernity* (Stroud: Sutton Publishing, 1998), pp. 203–24

Newman, S., *Embodied History: The Lives of the Poor in Early Philadelphia* (Philadelphia: University of Pennsylvania Press, 2003)

Noakes, L., *War and the British: Gender, Memory and National Identity* (London: I.B. Tauris, 1998)

Osbourne, M.A., 'Acclimatizing the world: a history of the paradigmatic colonial science', *Osiris*, 2nd series, 15: Nature and Empire (2000), 135–51

Park, R.J., 'Biological thought, athletics and the formation of a "man of character", 1830–1900', in J.A. Mangan and J. Walvin (eds.), *Manliness and Morality, Middle-class Masculinity in Britain and America, 1800–1940* (New York: St Martin's Press, 1987), pp. 7–34

Bibliography

Peers, D.M., 'Soldiers, surgeons and the campaigns to combat sexually transmitted diseases in Colonial India, 1805–1860', *Medical History* 42:2 (1998), 137–60

Peniston-Bird, C., 'Classifying the body in the Second World War: British men in and out of uniform', *Body and Society* 9:4 (2003), 31–48

Perks, R. and A. Thomson, 'Critical developments: introduction', in R. Perks and A. Thomson (eds.), *The Oral History Reader* (London: Routledge, 1998), pp. 1–7

Portelli, A., *They Say in Harlan County: An Oral History* (Oxford: Oxford University Press, 2010)

Porter, D., '"Enemies of the race": biologism, environmentalism, and public health in Edwardian England', *Victorian Studies* 34 (1991), 159–78

Porter, D., *Health, Civilization and the State: A History of Public Health from Ancient to Modern Times* (London: Routledge, 1999)

Post, S.G. (ed.), *Encyclopedia of Bioethics* (London: Macmillan Reference, 3rd edn, 2004)

Pringle, J.W.S., 'Effects of World War II on the development of knowledge in the biological sciences', *Proclamation of the Royal Society of London* A:342 (1975), 537–48

Rabinbach, A. *The Human Motor: Energy, Fatigue and the origins of Modernity* (New York: Basic Books, 1990)

Ramanna, M., 'Perceptions of sanitation and medicine in Bombay, 1914–1918', in H. Fischer-Tine and M. Mann (eds.), *Colonialism as Civilizing Mission: Cultural Ideology in British India* (London: Anthem Press, 2004), pp. 205–24

Rose, S.O., *Which People's War? National Identity and Citizenship in Wartime Britain, 1939–1945* (Oxford: Oxford University Press, 2003)

Rothman, D., 'Human experimentation: history', in S.G. Post (ed.), *Encyclopedia of Bioethics* (London: Macmillan Reference, 3rd edn, 2004), pp. 2316–26

Scarry, E., *The Body in Pain: The Making and Unmaking of the World* (Oxford; Oxford University Press, 1985)

Scheper-Hughes, N. and L. Waquant (eds.), *Commodifying Bodies* (London: Sage, 2003)

Schmidt, U., 'Cold war at Porton Down: informed consent in Britain's biological and chemical warfare experiments', *Cambridge Quarterly of Healthcare Ethics* 15:4 (2006), 366–80

Scott, J.C., *Domination and the Arts of Resistance: The Hidden Transcript* (New Haven, CT: Yale University Press, 1990)

Shephard, B., *A War of Nerves: Soldiers and Psychiatrists, 1914–1994* (London: Pimlico, 2002)

Shilling, C., *The Body and Social Theory* (London: Sage, 2nd edn, 2003)

Showalter, E., *Hystories: Hysterical Epidemics and Modern Culture* (London: Picador, 1997)

Bibliography

Silbey, D., 'Bodies and cultures collide: enlistment, the medical exam, and the British working class, 1914–1916', *Social History of Medicine* 17:1 (2004), 61–76

Smith, F.B., 'The Contagious Diseases Acts reconsidered', *Social History of Medicine* 3:2 (1990), 197–215

Smith, H.L., 'The womanpower problem in Britain during the Second World War', *Historical Journal* 27:4 (1984), 925–45

Smith, S. and J. Watson, *Reading Autobiography* (Minneapolis: University of Minnesota Press, 2001)

Springhall, J., 'Building character in the British boy: the attempt to extend Christian manliness to working-class adolescents', in J.A. Mangan and J. Walvin (eds.), *Manliness and Morality: Middle-class Masculinity in Britain and America* (New York: St Martin's Press, 1987), pp. 52–74

Sturdy, S., 'War as experiment: physiology, innovation and administration in Britain, 1914–1918', in R. Cooter, M. Harrison and S. Sturdy (eds.), *War, Medicine and Modernity* (Stroud: Sutton Publishing, 1998), pp. 65–84

Summerfield, P., 'Mass Observation: social research or social movement?', *Journal of Contemporary History* 20:3 (1985), 439–52

Summerfield, P., *Reconstructing Women's Wartime Lives: Discourse and Subjectivity in Oral Histories of the Second World War* (Manchester: Manchester University Press, 1998)

Summerfield, P., *Women Workers in the Second World War: Production and Patriarchy in Conflict* (London: Croom Helm, 1984)

Summerfield, P. and A. Carden-Coyne, *Contesting Home Defence: Men, Women and the Home Guard in the Second World War* (Manchester: Manchester University Press, 2007)

Talwar Oldenburg, V., *The Making of Colonial Lucknow* (Oxford: Oxford University Press, 1989)

Terry, J., 'Anxious slippages between "us" and "them": a brief history of the scientific search for homosexual bodies', in J. Terry and J. Urla (eds.), *Deviant Bodies: Critical Perspectives on Difference in Science and Popular Culture* (Bloomington: Indiana University Press, 1995) pp. 129–43

Terry, J. and J. Urla (eds.) *Deviant Bodies: Critical Perspectives on Difference in Science and Popular Culture* (Bloomington: Indiana University Press, 1995)

Thompson, E.P., *The Making of the English Working Class* (London: Gollancz, 1963)

Thompson, E.P., 'The moral economy of the English crowd in the eighteenth century', *Past and Present* 50:1 (1971), 76–136

Thompson, P., 'Introduction', in J. Bornat, R. Perks, P. Thompson and J. Walmsley (eds.), *Oral History, Health and Welfare* (London: Routledge, 2000), pp. 1–20

Turner, B., *Regulating Bodies: Essays in Medical Sociology* (London: Routledge, 1992)

Bibliography

Turner, B., *The Body and Society: Explorations in Social Theory* (London: Sage, 2nd edn, 1996)

Vernon, J., *Hunger: A Modern History* (Cambridge, MA: University of Harvard Press, 2007)

Vickers, E. 'The good fellow: negotiation, remembrance and recollection – homosexuality in the British armed forces, 1939-1945', in D. Herzog (ed.), *Brutality and Desire: War and Sexuality in Europe's Twentieth Century* (Basingstoke: Palgrave, 2009), pp. 109-34

Waitzkin, H., 'A critical theory of medical discourse: ideology, social control and the processing of social context in medical encounters', *Journal of Health and Social Behaviour* 30: 2 (1989), 220-39

Waldron, H.A., 'Occupational health during the Second World War: hope deferred or hope abandoned?', *Medical History* 41:2 (1997), 197-212

Walmsley, J. and D. Atkinson, 'Oral history and the history of learning disability', in J. Bornat, R. Perks, P. Thompson and J. Walmsley (eds.), *Oral History, Health and Welfare* (London: Routledge, 2000), pp. 181-206

Walvin, J., 'Symbols of moral superiority: slavery, sport and the changing world order, 1800-1940', in J.A. Mangan and J. Walvin (eds.), *Manliness and Morality, Middle-class Masculinity in Britain and America, 1800-1940* (New York: St Martin's Press, 1987), pp. 242-60

Wavell, A.P., *The Good Soldier* (London: Macmillan, 1948)

Webster, C., 'Healthy or hungry thirties', *History Workshop Journal* 13:1 (1982), 110-29

Weeks, J., *Sex, Politics and Society: The Regulation of Society since 1800* (London: Longman, 1981)

Wessley, S., 'Malingering in historical perspectives', in P.W. Halligan, C. Bass and D. Oakley (eds.), *Malingering and Illness Deception* (Oxford: Oxford University Press, 2003), pp. 31-41

Whiting, C. and E. Taylor, *The Fighting Tykes: An Informal History of the Yorkshire Regiments in the Second World War* (Barnsley: Pen and Sword Military, 2008)

Winter, J.M., 'Military fitness and civilian health in Britain during the First World War', *Journal of Contemporary History* 15:2 (1980), 211-44

Winter, J.M., *The Great War and the British People* (Basingstoke: Palgrave, 2003)

Woodward, R., 'Locating military masculinities: space, place and the formation of gender identity in the British Army', in P.R. Higate (ed.), *Military Masculinities, Identity and the State* (Westport, CT: Praeger, 2003), pp. 43-56

Woodward, R. and N. Jennings, 'Soldiers' bodies and the contemporary British military memoir', in K. McSorely (ed.), *War and the Body* (London: Routledge, 2013), pp. 152-64

Woodward, R. and T. Winter, *Sexing the Soldier: The Politics of Gender and the Contemporary British Army* (London: Routledge, 2007)

Zweiniger-Bargielowska, I., 'Building a British superman: physical culture in inter-war Britain', *Journal of Contemporary History* 41:4 (2006), 595-610

Bibliography

Zweiniger-Bargielowska, I., *Managing the Body: Beauty, Health and Fitness in Britain, 1880–1939* (Oxford: Oxford University Press, 2010)

Zweiniger-Bargielowska, I., 'Raising a nation of "good animals": the new health society and health education campaigns in interwar Britain', *Social History of Medicine* 20:1 (2007), 73–89

Internet sources

Historical Survey of the Porton Down Service Volunteer Programme, 1939–1989 (2001), http://webarchive.nationalarchives.gov.uk/20060715135118/http://www.mod.uk/DefenceInternet/AboutDefence/Issues/HistoricalSurveyOfThePortonDownServceVounteerProgramme19391989.htm (accessed November 2013)

BBC WW2 People's War Archive

A1109161, Dennis March, 14 July 2003, www.bbc.co.uk/history/ww2peopleswar/stories/61/a1109161.shtml (accessed November 2013)

A2427491, James Franks, 15 March 2004, www.bbcattic.org/ww2peopleswar/stories/91/a2427491.shtml (accessed November 2013)

A2669222, Harold Pollins, 26 May 2004, www.bbc.co.uk/history/ww2peopleswar/stories/22/a2669222.shtml (accessed November 2013)

A2772489, Harry Blood, 23 June 2004, www.bbc.co.uk/history/ww2peopleswar/stories/89/a2772489.shtml (accessed November 2013)

A3331577, Percy Bowpitt, 26 November 2004, www.bbc.co.uk/ww2peopleswar/stories/77/a3331577.shtml (accessed November 2013).

A5610485, Eric Middleton, 8 September 2005, www.bbc.co.uk/history/ww2peopleswar/stories/85/a5610485.shtml (accessed November 2013)

A6006359, L.W.A. Lyons, 3 October 2005, www.bbc.co.uk/ww2peopleswar/stories/59/a6006359.shtml (accessed November 2013)

A7747725, E. Owen Proctor, 13 December 2005, www.bbc.co.uk/history/ww2peopleswar/stories/25/a7747725.shtml (accessed November 2013)

Index

absence without leave 71, 137, 138, 144
acclimatisation 119, 145
Adam, Ronald 57
air force 5, 63, 102, 170
alcohol 75, 77, 137, 138, 142, 143
analeptics
  Benzedrine 92–3
  Methedrine 93
Army Bureau of Current Affairs 132
Army Medical Services 14, 120, 137
assault course 64, 105, 108
Auchinleck, C.J.E. 144

basic physical training 63–4
battle training 64, 187
'Blighty' wound 139
blood transfusion 97, 99
bodily size
  height and weight 4, 27, 34, 36–7, 45
bonding 12, 77
Bourdieu, Pierre 4–5
Bourke, Joanna 12, 16, 72, 154
British Expeditionary Force 125, 140
bromide 58
brothel 124, 126, 135, 142, 146
Brown, Ernest, Minister of Labour (1935–40) 28, 35, 36
burial regulations 20, 155, 161

Burje, E.T. 117–18, 121
Burma 7, 122, 159, 162, 169

casualty clearing station 158, 160
Chemical Defence Experimental Establishment 18, 95, 109
  see also Porton Down
cholera 119
civilian medical boards 27, 33, 39
class 4, 11, 12, 17, 37, 47, 79, 128, 174
comradeship 17, 66, 127
condoms 128
'conduct unbecoming an Officer and a Gentleman' 127
confined to barracks 61–2, 68, 76, 186
Connell, R.W. 3, 5
conscientious objector 26, 100–1
corpse 161, 167–70, 185
court martial 137–8, 141, 143–4
cowardice 138
Cowell, E.M. 125, 127, 130, 133
Crang, Jeremy 8, 32
cross-dressing 80
Cullumbine, H. 108

Dawson, G. 12
D-Day 160, 162–3, 168
death penalty 144–5
Defence Regulation 33B 125
demobilisation 15, 155

Index

desertion 71, 82, 137–8, 140–1, 144–5, 188
diet 3, 9–10, 12, 55–7
  see also nutrition
disabilities 9, 20, 30–1, 44–5, 119, 172
disability pension 14, 20, 27, 170, 173, 175–6, 178
'disgraceful conduct' 127–9
drill 1, 5, 18, 62, 65–6, 68, 81, 93, 130, 187, 189
drunkenness 75, 78, 137–8, 140, 146
  see also alcohol

Egypt 119, 122, 126, 129–30, 132, 141, 159
El Alamein 122–3
Elliot, Walter, Minister of Health (1938–40) 35

field ambulance 158, 167
field dressing station 158, 160
field ration scales 120
First World War 11, 26, 28, 32, 35, 38, 39, 47, 66, 72, 90, 95, 109, 127, 128, 144, 156, 161, 176, 185
Fleming, Alexander 98
Florey, (Sir) Howard 98
Foucault, Michel 2–3
French, David 8, 32, 143

games 13, 64, 66, 69, 93, 119
gender 3, 4, 12–13, 17, 125
  see also masculinity
General Service Scheme 8, 53
Goffman, Erving 4–5
Graves Service 161

haircut 1, 54, 61
Harrison, Mark 9, 57, 116, 127, 142, 145
health education 19, 131–3, 146
Higate, Paul 5–6
Holling, H.E. 105, 107

homosexuality 59–60, 76
Hong Kong 130, 140
hygiene
  Army Hygiene Service 122, 145
  Army School of 131
  Handbook of Military Hygiene 116, 118, 120
  hygiene officers 119, 123, 132
  Middle East School of 131
  organisation of 81, 123

India 116, 124, 139, 164
Industrial Fatigue Research Board 32, 94
industrial health 10–11, 35, 186
Industrial Health Research Board 11, 32, 94
infantry 5, 30, 53, 54, 61, 62, 67, 70, 71, 73, 75, 78, 138, 139, 158, 168, 172
informed consent 101
inoculation 119, 184
Interdepartmental Committee on Physical Deterioration 10
Inter-Service Topographical Department 121
interwar 11, 13, 32, 35, 56, 66, 94, 144

kit inspections 61

Leder, Drew 5, 16
Lees, Robert 126–8, 132, 137
Lines of Communication 30–1, 158, 161
love letters 76

Mackeith, S.A. 130, 135, 142
Mackworth, N.H. 96–7, 105, 108
malaria 100, 117, 121, 123, 130, 132–3, 139, 142
malingering 188, 43, 47, 72–4, 82, 128, 138–9
Mareth Line 164

Index

masculinity 3, 6, 12–13, 189, 190
Mass Observation 14–15, 45, 58, 73
masturbation 58–9
medical grades 28–30
Medical Research Council 18, 32, 37, 90
Mellanby (Sir) Kenneth 100, 106
Middle East 93, 98, 119, 121, 124, 126, 131–3, 135, 137, 142
Military Personnel Research Committee 91, 94, 105, 107, 109, 186
Military Training Act, 1939 26, 35, 37
militia 29, 36–7, 46–7
Ministry of Defence 102–4, 109
Ministry of Health 10
Ministry of Labour and National Service 10–11, 14, 26–7
Ministry of Pensions 100, 170, 172, 175–6, 178
Monte Cassino 133, 141, 169
morale 66, 70, 73, 127, 141, 145, 155, 161, 163
motion sickness 91
MRC *see* Medical Research Council
mustard gas 95–6, 107, 109

National Insurance Act, 1946 174
National Service Acts 1, 26
navy 63, 102, 160, 170
NCO *see* Non-Commissioned Officer
neurasthenia 117, 156
Non-Commissioned Officer 73, 79, 130
Normandy 135, 137–8, 158, 160, 163, 165–6, 169
  see also D-Day
North Africa 92, 94, 98, 116, 130, 133, 135, 143, 154, 159, 167–9
nutrition 55–7, 60, 120

officer–man relations 8, 66
oral history 2, 15–16, 103

pain 16, 159–60, 164–6, 177, 190
penicillin 98–100, 107, 124, 187
physical culture 10–11, 13
Physical Development Centres 57
Pioneer Corps 95
Porton Down 18, 95, 100–1, 103–6, 108–9
promiscuity 127–8, 135, 142, 146
propaganda 19, 132, 135, 142, 146, 185
prophylactics 19, 128, 137, 189
prostitutes 124–7, 147
psychiatric centres 157
psychiatric testing 38–40
public health 10–12, 35–7, 46
punishments 61, 68, 130, 143–4

queer culture 79

RAF *see* air force
radiography 39
  see also x-ray
RAMC *see* Royal Army Medical Corps
rationalisation 3, 9, 11, 32, 186
recruiting office 26, 41
recruiting regulations 26
recruiting standards 36
regimental aid post 139, 158–9
Regimental Police 61, 71, 74
route march 65, 68, 70
Royal Armoured Corps 96, 109
Royal Army Medical Corps 43, 54, 58, 69, 72–4, 92, 97, 108, 160, 167
Royal Artillery 57, 138, 140, 169
Royal Engineers 75, 103
rum ration 156

sanitation 116, 131–2, 146
scabies 100, 106, 129, 132
scientific management 11
self-inflicted wound 82, 138, 143, 188
  see also 'Blighty' wound

sexual behaviour 13, 19, 58, 60, 75, 81, 125–8, 136–7, 188
　see also homosexuality; promiscuity
shaving 55
Shilling, Chris 3, 167
sick parade 73
smoking 55, 163
sport 18, 66–7, 69, 126
　see also games
stretcher bearer 74, 159–60, 167
suicide 74–5, 188
Summerfield, Penny 15–16
sunbathing 118–19

Taylor, Frederick 11
tetanus 119
triage 159
tuberculosis 38–9, 44, 47, 139
Turner, Bryan 3, 14

typhoid 119–20

uniform 1, 13, 54, 61–2, 168

vaccination 120, 145, 184
venereal disease 58, 124–8, 130, 134–5, 143, 146, 185

War Office 14, 56, 90, 100, 102, 131
Watts, John 159–60
Western Desert 116, 120, 122, 124
Workmen's Compensation 72, 173

x-ray 38–9, 139
　see also radiography

yellow fever 119

Zweiniger-Bargielowska, Ina 10, 13

EU authorised representative for GPSR:
Easy Access System Europe, Mustamäe tee 50,
10621 Tallinn, Estonia
gpsr.requests@easproject.com

www.ingramcontent.com/pod-product-compliance
Ingram Content Group UK Ltd.
Pitfield, Milton Keynes, MK11 3LW, UK
UKHW021830210426